Geology and Astronomy

Geology and Astronomy

Charles Kovacs

Edited by Howard Copland (geology)
and Christian Maclean (astronomy)

First published in volume form in 2011
Fourth printing 2019

British Library CIP data available
ISBN 978-086315-807-0
Printed in Great Britain by Bell & Bain, Ltd.

Contents

Foreword

This book contains the lesson notes of Charles Kovacs, made when he was a class teacher of Class 6 pupils in Edinburgh. To some extent they have been revised and updated. In the geology notes the local area – in this case Edinburgh – is emphasised, and this is quite deliberate, as it serves as an example of relating what has been learned to the familiar, local area. One might expect lessons on geology to start from a description of Plate Tectonics — the great discovery arrived at during the 1960s. But this is the point — modern geology had existed for nearly 200 years and was a very advanced science before these discoveries could be made and appreciated. It is all too easy to give children 'ready-made' conclusions; conclusions which, in reality, evolved only gradually. The unfortunate result can then be that the conclusions are indeed accepted, but accepted as somewhat undigested dogma, not truly understood. The consideration of Plate Tectonics might come much better in the ninth grade, when time can be spent considering the diverse evidence — the 'detective story' which, eventually, led to such astonishing conclusions about the structure of the Earth.

In astronomy too the approach is to start from observations of sun, moon and stars — phenomena which have been experienced and understood — without special equipment or theories — for millennia. Once again, the upper grades are a better place to consider how the more recent concepts evolved. The later chapters on discoveries with the telescope are the most difficult to keep up to date, as new discoveries are continually being made, both with more powerful telescopes and with space probes.

We hope that these notes can serve as a stimulus to the creativity of individual teachers, rather than be followed as a scheme. This would be in keeping with the wishes of Charles Kovacs.

Howard Copland

Geology

1

The Children of the Earth

In the winter it can get very cold, but no matter how cold it can get here in our part of the world, there are places where it can get much, much colder. In which direction would we travel to find lands where it gets much colder than here? Would it be east, where the sun rises? South, where the sun stands highest at noon? West, where the sun sets? Or north, the direction in which we never see the sun moving across the sky? To find the colder lands we would travel north.

Travelling north we would come to lands which are much colder than ours; and the further we go north the colder it would get, until we come to parts of the world which are so cold that the ice and snow never melt completely and the ground is always frozen — summer and winter. Think of Greenland, Northern Canada, Alaska or Northern Russia. These are the lands around the North Pole, the polar region, the region of everlasting ice and snow.

On such a journey to the frozen north we would discover something else. We would notice something about the plants. In our part of the world there are two kinds of trees. First the leafy trees which, every winter, shed their leaves and stand bare. Then there are the trees that have only green "needles" for leaves; in botany we call them conifers or evergreens. Here, in our part of the world, we have these two kinds of trees, but going further north the leafy trees get fewer and fewer. There you would hardly see anything but the dark green of pines, firs and larches; the trees which have green needles as leaves.

Can you guess why this is so, why the trees with broad leaves get fewer and fewer the further we go north? Because a broad leaf tree needs more sun and cannot stand being frozen. Where there is less sunlight and it is less warm the trees with narrow leaves, or needles, can live better than the others. One could say that the leaves "shrink" and become needles as we go further north. Not only the leaves shrink as you go north; the trees themselves get smaller. The pines, firs and larches that grow far to the north of us are like dwarfs compared with our trees; they would barely reach to your shoulders. In the regions where these dwarf trees grow there is still a kind of summer. The snow *does* melt and then plants and flowers grow and have lovely blossoms; but they all have only tiny, tiny stalks, much shorter stalks than our flowers here. If we go still further north we come to the regions where nothing grows at all — not even dwarf trees or tiny flowers. These are the regions near the North Pole, the regions of everlasting ice.

But how would things change if we travelled further and further south — in the direction where the sun stands at its highest at noon? We know of course that it would get warmer and warmer. We would come to lands where people never have winter. How do the plants change as we go further south? Going south the needle trees become fewer and fewer, as they don't like too much sunlight, and the leafy trees increase; we come to trees with very large leaves indeed, like the palm trees. Think how long and broad and thick the palm leaves are. The trees also get taller and the other plants, the flowers, shoot up too. They have long stalks, long leaves and large blossoms in the hot lands of the tropics.

So you see how the whole earth changes from north to south. In the south, in the tropics, it is always summer — it is sunny and hot and we find tall trees, long stalks, big leaves and large flowers. The further north we go the trees are smaller, the leaves shrink, the stalks get shorter and, in the end, we come to lands where it is always winter and there is always ice and snow. Of course if you go beyond the equator, as far south as you can go, you finally end up at the South Pole and it is again extremely cold.

What I have just described is about the whole earth. The whole earth changes like this from south to north, or from equator to pole. But think of a very high mountain. Let us take the highest mountains there are in the world; the Himalayas. At the foot of the Himalayas you have a hot climate, the hot weather of India. You would see tall, broad-leaved flowers and trees. But as you go up into the mountains the air gets cooler and cooler, and the trees and plants become smaller. At a certain height you might think you are in Scotland — there are pines, larches, there is even heather, but there are also oak trees and beech trees. You go still higher and soon there are only needle trees and they become smaller. There are mountain flowers, like gentian, with short stalks. If you go still higher up there is, again, everlasting snow and ice. The summits of such high mountains are just like the polar regions and nothing can grow in the frozen, snow-covered heights.

Every high mountain is like the whole earth. Just as children often look like their father or their mother, so the high mountains of the earth are children of the earth, and they have a likeness to their mother, the earth.

2

The Story that the Mountains Tell

The hills and mountains of Scotland are very beautiful, but even our highest mountains in Scotland, the Cairngorms and Ben Nevis, cannot give you the kind of feeling that you have when you stand before the really high mountains of the world. If you have never seen the Alps before and are in Switzerland for the first time it can happen that you look up into the sky and think, "That's a strangely shaped white cloud up there." But when you look again, you see that it is not a cloud at all, but a range of snow-white mountain peaks, towering in the sky.

These mighty, powerful, majestic giants, reaching up into the clouds, leave you feeling something like awe before that power and greatness. But looking at these towering peaks you can also feel how immeasurably old these giants are. They have stood there for millions of years and they will stand there for millions of years to come. If these mighty peaks could speak they would tell us the life story of earth itself. We walk on the earth, we build our houses and cities on the earth and we use stones taken from the earth for our buildings, but what do we know about the story of the earth? The ancient, mighty mountains can tell us something about the story of the earth. Let us see what the mountains can tell us.

The first thing is that the great mountains of the world do not stand alone like proud giants, they are mostly to be found in groups or in long rows. These rows of mountains sometimes curve across the face of the earth for hundreds or even thousands of miles, and are called mountain chains or mountain ranges. The Alps are a mountain range, so are the Urals, and if you look at a map you will find that there are many, many more of these ranges. You can see that the Alps are really part of a much longer range that curves its way far to the east.

The next thing is that they are extremely old, unimaginably old, but — and this may sound strange — the mountains are not all of the same age; there are young and old mountains. The Alps are young and the Urals are old. Of course, even a "young" mountain at which you may be looking is much older than anything you can think of, but it is still young compared with another mountain.

Now we will compare the earth with the moon, the earth's companion in space. Astronauts have travelled to the moon and scientists have studied the rocks and mountains there. There are many great mountains on the moon, but they are all very old indeed — and they have hardly changed in the millions and millions of years since they were first created. The moon is beautiful, but its surface is like a barren desert, forever dead and unchanging; there are no young mountains at all, everything is very ancient.

Here on earth the young mountains are usually the biggest and tallest, with their jagged snowy peaks reaching far up above the clouds. Old mountains are not so big, although, when they were young, they too were just as tall as young mountains — like the Alps — are now. The old mountains have changed through time, they have been worn down and "rounded off."

So mountains tell us that the earth is not a dead place like the moon, it is active; old ranges are broken down and new ones are created. Always, somewhere on earth, there is destruction going on, but somewhere else there will always be new creation. Our earth is not just like some big lump of stone, it is a living, changing place — even when it comes to the seemingly lifeless rocks and mountains.

So that is the third thing: young mountains appear, old mountains wear down, and we learn that the earth is a place where great changes are always happening. And the way that mountains form in ranges tells us that the changes do not happen haphazardly — there is a pattern to them.

3

Young and Old Rocks: Granite

There are young mountains and old mountains but — and this is not the same thing — there are also young rocks and old rocks. Old mountains are made of old rocks and so you might think that young mountains must be made of young rocks, but this is not necessarily so. When nature makes a new mountain it uses rocks that are already there, just as a builder can use blocks of old stone to make a new house. In nature it is normal to recycle everything, and a young mountain range will be made of many different ages of older rocks. So there are young rocks, there are very ancient rocks and there is everything in between. What, then, are the oldest rocks?

To find the oldest rocks you need to look deep down into the earth. The oldest rock of all, the rock which lies under all the land and mountains of the earth under all our lakes and land, fields and forests, cities and roads — that kind of rock is a light-coloured rock called *granite*. Granite lies deep within all the continents of the earth. Beneath the soil on which we walk there may be clay and beneath the clay there may be limestone or sandstone and beneath that there may be something else — but if we go deep enough we shall always find granite. (Later we shall look at what is below the granite.)

But granite is not *only* to be found deep down, it can sometimes be found around us or at great heights. The Alps are partly limestone and other kinds of rock, but their highest peaks are of granite. Granite can reach from the deepest depth below to the highest heights above. We have granite in Scotland also: our Highlands, the Cairngorms for instance, are granite mountains. If you walk on this granite you walk on very old rock, you walk on rock that reaches deep into the earth and you walk on something that belongs to the very beginning of our earth.

There is a beautiful legend about granite.*

God wanted to create the strong, solid stone and rock upon which man should walk firmly through life. He turned to his helpers, the spirits and angels who serve him and he said, "Bring to me the gifts which you have so that the first of all rocks may be made."

Now there were three groups of angels around God. The first group were the angels of wisdom. The highest of the angels of wisdom came forward and he brought God the Father a stone that was as clear as water — it was transparent. The Angel said, "You, Father have given us the light of wise thought. This is the stone which is like the light of wisdom and the thought of man shall be like this shining crystal."

Then came the second group, these were angels of strength and power. The highest of these angels came before God and carried in his right hand a black stone, and in his left hand a white stone. The angel said, "These two stones, black and white, are the stones of strength. They will give man energy and strength so that his wise thoughts will lead him to deeds."

The third group were the angels of warmth and love. The highest of them brought a green stone and a red stone. He said, "In these stones we have put the warmth of our hearts. These stones can have many forms and will serve man in many ways."

And from these three gifts of the angels God made the first, the oldest of all the rocks, granite — from the gift of light, the gift of strength and the gift of warmth.

There are many varieties of granite, but it always is a mix-

* From *Erziehungskunst,* December 1952

ture of these three things: a clear, transparent stone called *quartz*, a black or white stone called *mica* (which glistens and sparkles in the light), and lastly a pinkish, white or greenish stone called *feldspar* (which gives granite its colour). As we shall find out later, the best soil for farming comes from feldspar. The bread we eat comes from feldspar earth.

The mighty granite rock, the oldest of rocks, is made up of three parts: quartz, the gift of *light*, mica, the gift of *strength* and feldspar, the gift of *warmth*.

4

The First Rocks

When a house is built the first thing is that a foundation is laid for the house. The foundation carries the whole weight of the house. Granite, the oldest of the rocks, is the foundation, the mighty giant that carries all other stones, rocks, earth and seas on its back. Beneath the oceans of the world, it is slightly different: the rock that supports the great oceans is a dark rock called *basalt*. Basalt is a relative of granite; you could say that it is like a younger brother. Basalt is dark in colour while granite is light, because basalt contains more iron and less of the clear quartz than granite. Iron is to be found in the basalt which forms the rocky bed of the great oceans. So the continents are supported on light-coloured granite and the oceans on dark-coloured basalt. Together these two "brothers" make up the foundations of the earth.

When you look at a granite mountain — in the mighty Alps, or the Highlands of Scotland, or even at small pieces of granite — you are looking at something so unbelievably old, so ancient, something that came into existence so long ago, that no one can know for certain how it happened. So there are two different ways of explaining how granite came to be.

Before the first explanation I want to tell you of something I saw as a child in Austria. During the holidays my parents used to take us to a little town called Baden, which means *bath*. Why was it called that? Because this town had a spring of water that came from deep down in the earth and this spring of water was hot. It was nearly boiling hot as it came from the earth. It was not heated by people but by the earth itself. It also had a peculiar smell, like eggs going bad. To take a bath in this hot, smelly water was a very good cure for rheumatism and so people came from far and wide to take these baths. And there are hot springs in many other parts of the world: England, Iceland, New Zealand, America, Japan.

This water is heated deep down in the earth. When people dig deep mines for coal and for iron they find that the deeper a mine-shaft goes, the warmer it gets. In some mines the shafts go so deep that they need special cooling systems, or the men could not work there at all. We saw earlier that the higher we go the colder it gets; now we can see that the deeper down we go the hotter it gets.

In steel-works the furnaces produce a heat in which iron melts and flows like water, it becomes a white-hot liquid. If one could dig down deep enough it would be so hot that there would no longer be any hard, solid rock. Even rock would melt and be a red-hot liquid. In our time one would have to dig a shaft of about two thousand miles through the rock to get there and, of course, no one can do that. But some people who study these things say that in a far-away past, millions of years ago, you did not have to dig down to find this great heat in which even stones and rocks were flowing like water. They say that the surface of the earth, where we walk about now, was like that, was so hot that there were no rocks or stones. It was all a fiery-hot liquid. Now in time this surface of the earth began to cool down — at first only the outside, then a bit deeper down, and then still deeper down. As the outside surface slowly, very slowly, cooled down over thousands and thousands of years, it hardened — just as molten iron hardens when it cools down — and a hard skin, a hard crust formed on the surface of the earth. Now this first, hard, solid skin of the earth was granite.

That is one way of explaining how granite came to be; it is called the "hot-earth" theory. But there are people who think differently, they prefer the "cold-earth" theory. They say it is not surprising that it is terribly hot in the depths down there: if you press with your hand hard on the desk, it will get quite hot — and so if all the heavy mountains and oceans and rocks press down, it must get hotter and hotter the deeper down you go. But, they say, this does not mean that it was ever so hot on the surface. Now these people say that the earth is not just a great dead chunk but is more like a living being. We know how, for example, shellfish — crabs, lobsters, sea urchins, — form a hard shell around themselves, well perhaps the earth could have formed a hard shell around itself, and that shell is granite.

But the time when the first granite came into being is so terribly long ago that no one can say for certain how it was. Granite, that oldest of rocks, that mixture of three stones, is still keeping the secret of how it first came into existence.

5

Volcanic Rocks

Think of a very high granite mountain. Its peak is so high up that it is forever winter, covered with ice and snow. But the granite goes on under the earth, deeper and deeper down, so far down that it comes to the terrible heat below where the rocks and metals are burning hot. The granite mountain reaches from the terrible heat below to the terrible cold above. But we living beings, humans and animals and plants, we live on the surface of the earth, just between the two extremes — we live in between the terrible cold of the heights and the terrible heat of the depths. You see, life is always something that keeps to a middle path between the extremes of too much or too little. That is something to remember for our own life; that the middle path is the best.

In some parts of the earth's crust — not everywhere, only in certain places — the heat is so great only a few miles down that the rock has melted and become a red-hot thick liquid. This molten rock is called *magma* and it forms an underground magma chamber. The magma can remain there for long times, but it does not always stay down in the depths. At certain places and at times which no one can foretell this magma comes spurting up from the depths. And it is always an awe-inspiring thing when this happens. Many years ago, magma came spurting out at one place. It had such a terrific strength that it forced a path, like a long pipeline, through the rock, through the soil and everything on top, then came spurting out through a big hole in the ground. It was fiery-hot, but in the cool air on the surface the magma soon cooled down and hardened. It

A volcano forming a crater

hardened into stone, into rock. What remained was a little hill of hardened rock with a hole in the middle where the magma came out. The next time there was such a spurting of magma from the depths it did not force a new path; it came out where there was already a hole. So a new hill was formed on top of the old one, but the old hole still remained. The next time the same thing happened the hill grew and, in time, became a great mountain with a crater. The deep hole through which the eruptions come is called a *crater* — the Greek word for a mixing-bowl.

Now the Romans had a god who was a metal-smith called Vulcan. He made the weapons for the other gods. They said,

"As a human smith has a smithy where he heats iron until it is soft so that he can work on it, so this god has an enormous smithy, deep down in the earth. Magma comes from the smithy of this god Vulcan." The mountains formed by magma are called *volcanoes* after this god. So you see, a volcano is quite different from other mountains. A volcano is like a man who is quarrelsome, an ill-tempered person who does not get on with other people and so has to keep to himself; the mountain ranges on the other hand are like friends who stand together. Volcanoes are not formed by the slow process that forms mountain ranges but by sudden fiery outbursts called *eruptions* — quite a different story!

A volcanic eruption can be a fearful thing. First there is a deep rumbling noise from inside the earth, then a puff of steam, ash and smoke comes from the crater. Another deep rumbling comes and the earth is shaken for miles around. Then from the crater a fountain of fiery liquid erupts and runs down the mountain slope, like red-hot snakes. At the same time an enormous cloud of fiery smoke and ash forms above the volcano's peak. This cloud darkens the sky until it is dim as night, and a rain of ash falls down on the surrounding countryside for miles around. But this terrible, fiery-hot cloud can also come right down to earth, rolling down the mountain-side with unstoppable force and speed. When this happens it burns and destroys everything in its path; it can even destroy a whole town if it happens to be in the way.

The fiery stuff that comes from the crater and flows down like snakes is called *lava*. Lava is really the same as magma — it is called magma when it is beneath the earth and lava when it comes out on the surface. After a while the lava cools enough to become solid rock. Often, as it cools and hardens, it forms big wrinkles on its surface, looking like piles of rope; this is called "ropy lava." From lava, different varieties of rock are formed. It can happen that the magma does not reach the surface; then, because it is deep in the earth, it cools down slowly and becomes *basalt*, the dark rock which also makes up the crust of the earth beneath the oceans. Then there is *obsidian*, a kind of natural glass that forms when lava cools quickly — it looks like dark bottle-glass and can be found in many colours. The Native Americans had not discovered the use of iron before Europeans came, but they made razor-sharp knives and arrowheads out of obsidian. Just as sea water foams or beer froths, so the liquid red-hot lava can also make a bubbling froth and when that hardens it becomes *pumice*, the stone you use to rub ink-stains off your fingers. It is a strange kind of stone — full of bubbles and so light that it floats on water.

If a volcano erupts regularly it is called an *active* volcano. But a volcano can be quiet, without any eruptions for a hundred years or longer, and then it suddenly erupts again. A volcano that has not erupted for a long time but that — one day — will erupt again, is called *dormant*, which means "sleeping." A volcano that is completely dead and will never erupt again is called an *extinct* volcano. As it happens, around Edinburgh there are a good many of these ancient volcanoes — Arthur's Seat, the Castle Rock, Calton Hill and, farther to the east, Berwick Law and the Bass rock. They are all dead, extinct volcanoes — but once they were fire-spouting and active.

It is a strange thing to go up a volcano which is still active but, for the time being, not erupting — like Mount Vesuvius, near Naples in Italy. On the lower slopes of the mountain are fields and vineyards, for volcanic soil is very rich. You walk further up on bleak, desolate slopes, raised in heaps like the waves of the sea — it is all hardened lava. In many places you walk on a layer of clinker. Every now and then the ground under you trembles and a rumbling noise comes from the earth. It is a very tiring walk up but at long last you reach the top and you look down into the immense crater. It is like an enormous basin with sheer sides and you can see that it is made of layers of lava one on top of the other. In places there are wisps of steam and there is the constant sound of small stones falling down into the crater. You climb down a short way into the big crater of Vesuvius. The walls are hot to the touch: the heat of the earth. As you get down into the crater little lizards run away in different directions; they like the warmth in the crater.

Throughout history Vesuvius has erupted many times, but the most famous eruption was around two thousand years ago. About 10 km or 6 miles from Vesuvius (quite a long way!) lay the wealthy Roman town of Pompeii. In the summer of AD 79 a great eruption of the volcano threw up huge quantities of smoke and ash high into the sky. Day was turned into night and a heavy rain of ash came down until the town was completely buried. In our time it has been dug up again, revealing a lot about life in Roman times (it is said that the science of archaeology was born in the ruins of Pompeii). It is fascinating to visit Pompeii and walk around the streets of the ancient town.

In some other volcanoes — as on the Hawaiian Islands — you discover that there is almost continuous but much less violent activity. Inside the big crater there is liquid lava that spurts up in a red fountain — especially at night it is a spectacular sight. The fountain can be quite high, thirty metres (100 ft) or so and it throws out red-hot drops of lava that harden quickly once they are out and fall as "bombs." Again you feel the trembling of the earth and hear the rumbling below. Here and there you can see patches of yellow sulphur around the cracks where steam is coming out of the ground and there is a pungent smell in the air. When you see and hear and smell this, then you know that one does not have to dig deep down to reach the magma below the earth — in a volcano it is as if the inside of the earth comes up to the surface.

6

The Restless Earth

If you visit the seaside town of Pozzuoli (a suburb of Naples, not far from Vesuvius) you are at another place where you can see and feel that the activity of the earth comes near the surface. On the edge of town is the Solfatara. When you approach this place, you will probably smell it before you see it, for there is an unpleasant smell, like rotten eggs. The Solfatara is a large shallow crater full of light coloured sand, but in places there is steam hissing furiously from the ground, as if there were great kettles boiling just below the surface. Around the steaming vents are puddles of spitting, boiling mud and there are coloured deposits of sulphur and other minerals, most of them very poisonous. If someone throws down a ball of lighted paper, the ground in that place "answers" with its own smoke and steam; it can be so strong a reaction that you are surrounded by a cloud of it. At times it gets so active that it has to be closed off because the fumes are too dangerous. It almost feels as if the earth there is showing its anger.

We have looked at the great changes, the complete transformations, that have come about over long periods of time. The whole "face of the earth" is in slow but powerful movement. Not only the dry land is going through such movements which after thousands and thousands of years produce mountains and valleys. The sea-bed, the solid rock under the sea, rises in places and so the sea-bed that used to be underwater comes up and becomes dry land. And the opposite also happens: dry land sinks down becoming the sea-bed.

On the other side of Pozzuoli are the remains of an ancient

The so-called Temple of Serapis, Pozzuoli, was actually a market place

Roman building, the so-called Temple of Serapis. Not far from the sea there are three tall marble columns, still standing today where they have stood for about two thousand years. If you look more closely at these columns, there is something rather strange about them. About 4 metres (12 ft) from the base of the columns there is a band of discolouration and roughening of the marble. What can have caused this damage? The surprising answer is that the columns are pitted with holes made by a kind of shell fish which lives only in the sea. Now this shellfish could only have made its marks under the water, so the pillars must have been deep in the sea, and then came out of it again. It means that the sea bottom went down, then it rose again — and this may have happened more than once. At present the marks on the pillars are roughly 7 m (24 ft) above sea-level, — yet this is not fixed, there is always a slow change going on. But these up-and-down movements of the earth can also be sudden. In 1984 there were earthquakes in the area and afterwards it was found that the sea bottom at Pozzuoli had come up by nearly 2 m (6 ft), making the bay too shallow for big ships. So in this part of Italy the earth is more restless, rising and sinking faster than in other places.

There are other parts of Italy where earthquakes strike. In the year 1908 a terrible earthquake — perhaps the worst ever known in Europe — struck the city of Messina, an important seaport on Sicily. On December 28, only a few days after Christmas, the earthquake occurred in the early morning, when most people were still asleep in bed. There was a great shaking: the earth rose and fell like great waves of the sea, accompanied by deep rumblings. Within seconds the walls of houses, palaces, churches, came crashing down, crushing and burying thousands of people under them. But worse was still to come. An enormous wave from the sea came thundering in — a tsunami — higher than the houses, swept in from the shore, and many who had somehow survived the earthquake now died under the raging flood. At least 70 000 people died in that dark morning of terror.

Let us remember the slow, majestic movements of the earth that build the mountains, but let us also keep in mind how terrible these movements can be when they are fast and sudden.

7

Some Different Rocks

One could say that in granite you see the good temper of the earth, the love of the earth. In volcanic eruptions, in volcanoes, you see the bad temper of the earth, the anger of the earth. It may well be that the great, terrible eruptions and earthquakes — not the slight rumbling which goes on in some parts of the world all the time — are outbursts of bad temper from the earth over the evil things done by people on earth. The earth wants us to be like granite — wise, strong-willed, kind; but if there is much evil in the world then sooner or later the earth will burst out with bad temper, to show that it is annoyed.

Basalt, the dark-coloured relative of granite, sometimes shows another side of its nature. When you look at Arthur's Seat in Edinburgh you can see Samson's Ribs — great curving columns of rock that look like the ribs of an enormous giant. These basalt columns are not round, they have a *hexagonal* form — they have six flat faces, like giant crystals. Another place where you can see these basalt columns is on the little island of Staffa in the Inner Hebrides, in the famous Fingal's Cave that is like a vast natural cathedral. When the great musician Felix Mendelssohn visited Staffa he was so moved by the sight of the towering columns and by the strange echoes inside the cave that he wrote a wonderful piece of music — it is known as the *Fingal's Cave Overture.* Painters and poets have also been inspired by it. In Northern Ireland there is another natural wonder, the Giant's Causeway, that is also made of these hexagonal basalt columns. Long ago Fingal's Cave and the Giant's Causeway were connected; they were once part of a huge lava flow that happened millions of years ago. It is hard to imagine such a gigantic lava flow happening — no lava flow today is so huge.

The dark rock, basalt, is dark because it contains iron. But it also contains another metal called magnesium. We have iron in our blood, it is iron that gives our blood its red colour. But magnesium is just as important for the plants — it is magnesium that gives them their green colour and their ability to take in sunlight. Earlier we wondered what lies below the earth's crust, below the granite of the continental crust and below the basalt of the ocean crust. Sometimes this deep rock comes to the surface through volcanic activity, and it is also a rock that contains iron and magnesium, these essentials of life.

But there are many other substances in the rocks and there are many other different kinds of rocks. In the crust of the earth there is abundant feldspar; and feldspar contains *calcium*. We all have calcium in us because it is calcium that builds our bones and makes them strong. Calcium is also found in the shells of sea creatures, crabs, sea-urchins, molluscs. This brings us to a kind of rock that was made in a quite different way to granite or basalt.

Imagine a shallow tropical sea that existed long ago, many things live in it — fish, corals, shell-fish. Floating in this sea, making it slightly cloudy, are millions of very tiny creatures with beautiful white shells. When these tiny animals lived their lives and died, they sank to the bottom of the sea and their shells made a kind of chalky ooze. This went on for hundreds or thousands of years so that great heaps of these shells made a thick covering on the bottom of the sea. Over time these heaps were pressed down so much that they became a kind of rock called *chalk* which is a soft variety of *limestone.* So chalk and limestone begin life at the bottom of the sea.

The White Cliffs of Dover, the Downs and the Cheddar Caves in the South of England are all made of limestone. And limestone is a very widespread rock — it can be found making high mountains like the Jura — and it can be found deep under the ground over large areas of Europe and other parts of the world. But these hard limestone rocks were originally made by little sea creatures — you could say that limestone is a kind of animal-rock. So, although the limestone rocks we see now are lifeless, they were formed originally by living creatures.

We may not think it makes much difference to people whether they live on granite or limestone, but there is a difference. People living near granite feel more awake, feel more strength in their limbs, feel more active, and want to do things. But those living in limestone country feel more dreamy, and not so energetic. Some people prefer one, some the other.

8

More about Limestone

Just think of the different ways that rocks and mountains form. Limestone mountains, which are made of millions of tiny shells pressed together, have come about in a quite different way to volcanoes. Now they are both part of the hard crust of the earth, but volcanoes have their origin, their beginning in *fire* and limestone mountains have their origin in *water*. You remember the four elements: fire, air, water, earth; each has its part to play in forming the world around us.

The White Cliffs of Dover were originally formed at the bottom of the sea, but now they stand high over it — so how is limestone found far above sea level? As our chest goes up and down when we breathe, so the earth can sometimes rise slowly and then go down — but it may go down in one place and in another it may go up. In some places the sea-bed has risen up and become land (as we saw with the Temple of Serapis near Naples). The sea-shell rocks now stand on dry land as hills such as the Pennines (limestone) or the Downs (chalk). And this "breathing of the earth" happened not only here in Britain; large parts of Europe, America, Africa which are now dry land were at one time seas.

And limestone can be found much higher than the Pennines. It is an important part of the Alps and it is at the top of the highest mountains of all, the Himalaya. It seems strange that you can find the remains of ancient sea-shells high up above the clouds in these mountains. But there is another reason for this. When the earth makes a new, young mountain range, it uses rocks which are already there. Limestone was among the rocks which were already there — it was already quite old when these young mountains were formed — and so it was taken up in the great squeezing and folding that formed the Alps and Himalayas.

Limestone was formed in warm tropical seas, but it can be found in all the different climates of the world, not just in the tropics. The sea around Dover is certainly not tropical — so how did limestone come to form there at all? Not only can the land rise and fall above or below the sea, it also moves across the face of the earth. A land that was once freezing cold might now be a hot desert, or a place that was once a tropical sea might now be somewhere far to the north with a cold climate. As the continents slowly move on the face of the world, everything changes.

These rocks and mountains that came from water differ among themselves very much in hardness. It depends how hard they were pressed down and how well the little shells were stuck together. Chalk is the softest — it crumbles, as we know, very easily. Limestone is harder and does not crumble so easily. Marble, the beautiful white stone from which the Greeks and Romans made their statues and their gleaming white temples is also a kind limestone, but it has changed. During the slow movements of the earth, limestone can sometimes be pushed deep underground (some rocks are pushed up and others are pushed down). Down there in the earth the limestone was heated so much that later, when it came to the surface again, it had been transformed into beautiful crystalline marble. But limestone and marble are not always white, they can become mixed with other substances — sand, clay, plant-remains — and these mixtures cause many beautiful colours and patterns in the marble.

But all these rocks and mountains that came from the sea, even the hardest of them, are softer than granite or basalt. Wind and water can gnaw and nibble at them more easily than at the harder stones. That is why limestone rocks often take on strange shapes: they may look like castles or knights.

And limestone can, slowly, dissolve in rainwater. Rain soaks deep down into the ground and in the limestone regions of the world there are all kinds of underground streams and tunnels, fantastic caverns with stalagmites and stalactites, and even underground lakes. Some people, called *potholers*, love to spend their time exploring these strange and wonderful — but sometimes also dangerous — underground worlds.

Limestone is also very useful to us; it is used a lot as a building stone. The pyramids of Egypt and the Houses of Parliament in London are both made of limestone. It is also used in making important modern building materials: cement and concrete. Altogether limestone is one of our most essential building materials, and your own strong bones are really a kind of living limestone. You see how every kind of rock has its own story.

9

Coal

Limestone is a kind of "animal rock" because it is made of the shells of sea animals. You can also find some kinds of limestone which started out as ancient coral reefs. (Coral is also an animal, even though it looks a bit like a plant — with some of these ancient life-forms it is not always easy to tell plants from animals.) Is there also a kind of rock made from the remains of plants? Yes, there is, it is the black rock which we know as coal.

We burn coal in our power stations by the millions of tons because the heat given out by the burning coal can be changed into electricity — and we need this electric power to keep our modern world running. There are other ways of making electricity, but we still depend on coal to make the huge amounts of electricity that we need. But we have also learned that if we burn too much coal it is a bad thing — the burning of the coal changes the air: it increases the carbon dioxide and that causes changes to the climate. This is one of the problems of the present day and also of the future.

Of course plants are mainly green, so why is coal black? If you look at a compost heap, you will see that when plants have died and turn into compost, they first go brown and then black; by then they have become good "humus" for the soil. Or think of a piece of charcoal, it is made of wood which has been charred, but not completely burnt to ash. The black colour of charcoal is a substance called *carbon* — and all living things, both plant and animal, contain carbon. The best quality coal is nearly pure carbon, but this carbon was once part of living plants.

But what sort of plants were they and how did they become coal? Just as there was a time in the history of the earth when huge amounts of chalk were created, so there was an even earlier time — the *carboniferous* time — when huge amounts of coal were formed. (These were not the *only* times when limestone and coal were made, but they were the main ones.) In those days there were great forests, not just here or there, but covering large parts of the earth. These forests were not like the ones we know today, the plants that grew in them looked different. Today we have ferns and horsetails but in the carboniferous time there was a much greater variety of ferns and horsetails, many of them were big trees, and there were other strange trees, not much like anything we have today. Nowadays there are still some tree-ferns, but the trees we know so well — oak, ash, sycamore, birch — did not yet exist in those days. Not only that, but there were no flowering plants at all — they had not yet come into existence — there was not a single flower anywhere in the world. So the hot swampy forests of the carboniferous time would have seemed very strange to us.

How did these swampy, peaty forests become the coal-seams which we can dig up today? It is always difficult to be sure about what happened so long ago, but it seems that the sea-level changed many times. The sea rose until large parts of the forest were completely flooded and lay under water. On the sea-bed the plants died and lay in heaps, getting slowly covered by layers of mud or sand or shells. The plant remains were pushed down by the weight of the water and the layers of mud above them until they were squeezed together and formed a thick black layer, sandwiched by mud above and below. Eventually the sea-level fell again, leaving muddy, sandy land with no plants growing on it. But soon new seeds fell in the mud; plants began to grow and — in time — there was a whole new forest, just as big as the old one.

More time passed and the sea came up again, drowning the new forest and making a new layer of black plant-remains at the bottom of the sea. This slow rhythm of land, sea, land, sea, went on for millions of years until there were many layers, one above the other. You can find 40 seams of coal (or even more in some places) with sandstone or limestone in between. These repeated layers in the crust of the earth are a bit like the growth-rings in the wood of a tree, they are formed by the slow, rhythmic ebb and flow of life. And, like the growth rings, the layers are all slightly different; some are thinner and some are thicker. But there are two big differences.

Firstly, the layers of the rock are far bigger than tree-rings, 3 metres (10 feet) is not unusual and some are much thicker. Secondly, one tree-ring is made in one year, but the layers in the earth are made very slowly — thousands and thousands of years for each one. We are used to living with only two rhythms of planet earth — the day and the year — but there are other, much longer, rhythms too. There is an important one (precession) which lasts about 26000 years — you will hear about it when we study astronomy. There are other, even longer rhythms in the life of planet earth: 40000 years, 100000 years. It may be that these are the very slow rhythms which have left their mark in the coal measures.

There are also different kinds of coal, depending on how much the coal seam has been pressed down by the layers above. The coal which is found very deep in the earth has been pressed down hardest, it has also been heated most and it is the best kind of coal. It is called *anthracite* — it gives better heat and it burns cleanly, without much smoke — but it is difficult to get at it, and so it is scarce. Most coal is of a type known as *bituminous*; it has not been pressed down so much so it is softer — and it burns with a smoky flame. The softest coal is dark brown; it is called *lignite* and it is like peat which has been compressed.

Finally, why does coal contain so much power — why is it that we can use the heat and power of coal, but not any other kind of rock which we dig from the ground? It is because the plants in those ancient forests soaked up a little of the heat and power that streamed down on them from the sun. In coal some of the heat and power of the sun are "locked up." And when the coal is set on fire it is actually the warmth of the sun which is released from its long "enchantment" in the black rock.

It is really the same with each one of us — inside each one of us there are hidden powers, gifts or talents, which can be released if something "sets us on fire." All we need is to be open for the hidden powers and potential in ourselves and in others.

10

The Work of Water

We have heard how mountains and rocks came into existence, but nothing in the world stays as it is. Mountains are "born" and they "pass away" as we do. All things change; and even the most mighty and ancient mountains of the world change; and what changes them more than anything else is water.

Think of a big rock. Like all rocks, it is full of tiny cracks. In the rain the tiny cracks fill with water. When the cold winter comes the water in the cracks freezes and turns into ice. It's a strange thing, but when water turns into ice it expands, it grows bigger — and it does so with an enormous, unstoppable force. With such a force the crack is widened. In summer the ice melts, but sooner or later it freezes again and, after this has happened many times, the crack is so big that a bit of rock breaks off. You can often see scree in the mountains — a slope of loose rubble and stones. All these stones of the scree have been broken off from the big rocks by the freezing and thawing of water.

When the rain falls it runs down the slope in little runnels, and the little streams come together and form a bigger stream, and the bigger stream pushes the stones along — but all the time these stones are being rubbed against each other. As the water pushes the stones and rubs them against each other, the corners get knocked off and, eventually, they become round pebbles. That's what you can see in any stream. The smooth, round pebbles were once sharp-edged stones. They have been rounded off in the running water.

The water of the rushing stream keeps on rubbing the little

stones against each other — small bits are rubbed off — and so they get smaller and smaller until they become little grains; we call them sand. Sand is rock that has been ground smaller and smaller by the running water. All the time the water carries the pebbles, gravel and sand further downstream. The smaller the stones become, the more easily the water can sweep them along, so that upstream there will be more pebbles and downstream there will be more sand.

As the granite (for instance, from the Cairngorms) is broken down, the quartz and feldspar are separated. Quartz is harder than feldspar and so it does not break down so much, it becomes sand. The feldspar is less hard and the pieces of feldspar get smaller still until they are like grains of dust, and in the water these tiny grains of dust become something like mud. The proper name for this mud is clay. If you rub sand between your fingers it feels rough and gritty, but clay is so fine that it feels quite smooth.

So, the water breaks the rocks when it is ice. It sweeps the broken bits of rock downstream and they become gravel, then they become sand, then they become clay. Then the sand and the clay are taken further downstream by the river. As the river gets nearer to the sea it usually gets wider and it slows down. The slow-moving water does not carry the mud so well and a great part of the sand and clay is "deposited," that means it falls to the bed of the river and lies on the river banks. We have all seen sand and mud on the banks of a river. When the river gets to the sea, it brings sand and clay with it and a lot goes right out into the sea, finally landing on the sea bed.

When sand, clay or mud fall to the bottom of the sea or a lake and make a layer there, the layer is called *sediment*. As we know from the example of limestone, sediment can be pressed down and in time turn to rock. So rocks which form in this way below water are called *sedimentary* rocks. Limestone is one kind of sedimentary rock and now *sandstone* is another. Sandstone can form if the piles of sand are pressed down hard over long ages. But not all the sand is under the sea; by the action of the waves in the sea, some of it spreads along the shore; that's how you get sandy beaches.

All this has been going on for hundreds of thousands of years, the rivers making sand and clay. When it lies on dry land, plants grow on it and creatures burrow in it — they live and die year after year — and the top layer of sand and clay becomes soil. All the good soil in the world, the soil in which plants can grow strong, the soil which gives us all the crops we need for our food — all the soft fertile soil in the world is possible because of the sand and clay which the water has brought down from the mountains.

No plants could grow, no animal and no man could find food if the water did not nibble away at the mountains and if the water did not rub stones together until they became tiny grains of sand and clay. When you see the good soft earth in gardens and fields then think that this was once hard rock high up in the mountains. Think of the water which slowly nibbled at the mountains, washing them away, but by doing so gave us the good soil which we need for our food.

11

The Circulation of Water

Now we must get to know a bit more about the waters of the earth. We have all seen water boiling and — rising up from it — the clouds of hot white mist we call steam. If a kettle boils long enough all the water would boil away and go up in steam. If you hang your wet washing out on the line, the washing will dry; the water in it will also disappear, it will also rise as "steam," but such a fine steam that you can't see it. And if I put a drop of water here it will also dry up after a time — it will disappear. How? It also rises up as invisible steam. In Latin, *vapor* is a word for steam, and the disappearance of water is called *evaporation* — it evaporates. What makes the water evaporate when there is no fire under it? The warmth and light of the sun; the warmth from above. And now think of all the oceans and seas and lakes all over the world. The sun shines on all of them and from all the waters of the world a huge amount of invisible steam rises. Now imagine that the drops of water could speak and tell their story. They would say:

We were on a big wave in the great ocean and the shining rays of the sun came and with their powers they lifted us up. We rose higher and higher — we became light as air — and more and more of us came up and we all became a beautiful white cloud. Oh, it was lovely to float up high above our father, the ocean. But the wind came and carried our cloud away — we flew over lands, over fields and forests and mountains. Near the mountains it grew cold and our great white cloud turned dark — the cold air made all of us heavy — we got so heavy that we could no longer stay up and we all fell down as rain.

Some of us fell on the fields, some on forests and trees and flowers, but they all were glad to receive us. Many of us fell on mountain rocks where the rays of the sun lifted some of us up, others seeped into the rocks, and others again trickled over the rocks and where a few trickles met, we joined joyfully together and made brooks. The little brooks met — it was lovely to see the other drops again — and came together in a little stream that tumbled downwards over the rocks. Soon the little stream met another little stream, and another one, and we all joined together. Now a strong mountain stream rushed swirling down and met other mountain streams. And now together we became a rushing river. The river flowed on, and joined with another river.

Now we were a great wide river that flowed through the plains, through flat land. The river carried many ships with goods and passengers, and ferries going from one bank to another. Great bridges spanned it and great cities stood on the river banks. But on it flowed, and then we reached the sea, the ocean. Joyfully we came back from our long journey to our ocean father. But soon we shall be ready again for the next journey when the sun's rays lift us up to the sky.

That is the life-story of a river, the water which flows from the mountains to the sea. It came from the sea, it rose up from the sea as clouds, and came down as rain. It is really the sun which makes our rivers flow. Without the sun there would be no clouds, no rain and no river. The whole path which our drops have travelled from the sea back to the sea is like a circle that comes back to where it started — it is called the *circulation* of water or the water *cycle*. The blood in our bodies also circulates — it goes round and round — but the "sun" of our blood is the heart.

We saw that the streams and rivers slowly break down the mountains and grind the rocks into sand and clay, into soil. What really makes the good soil? It is the sun. The sun not only makes the plants grow; it has even made the soil from which they grow!

12

Winds

We have learned about earth and about water; now we must also learn about the air. We can't see the air; it is colourless. A room is full of air, but we can't see it. Things we can't see are real just the same. If the air in a room gets used up and we feel it's stuffy and want to have some fresh air by opening the window, then we can be sure that this air we can't see is quite real.

But there are ways of making the air visible. Water is also colourless, but we can colour it with paint. Quite often air can be coloured and become visible because something has been put into it. We all have seen coloured air quite often. I am talking about smoke — the smoke rising from a fire is nothing but air coloured by tiny bits of ash.

These bits of ash do not just rise by themselves. They rise because the air carries them, and the movement you see is nothing else but air rising upwards. We all know why the air from a fire goes up: it is hot. We know that steam rises up from boiling water, and air, when it gets hot, also rises. What will cold air do? It will sink. In a fireplace the hot air goes up through the chimney, and from below colder air comes and takes its place; that is what is called a draught. If there is no draught the fire will not burn. This process — hot air rising, cold air coming in below and taking its place — goes on in nature, and that is how we have winds.

Now let us take something a bit more complicated. Imagine it is summer and we are at the seaside on a really nice, sunny day. The sun shines on the land and also on the sea. By about lunch-time you will find that stones and even the top sand have become quite warm — but the water is cool, much cooler than the land. Earth and stones get warm much more quickly than water. But if the earth is warmer, the air above the earth also becomes warmer, and rises. Cool air then comes from the sea. So on a warm, still day, in the afternoon you will get a cool breeze coming from the sea.

At night at the seaside, the earth, stones and sand which became warm so quickly in the sun also lose their warmth quickly; but the water, which warms slowly, also cools down slowly. The water is still warm when the land is already cold. So, in the night, there will also be a breeze, but from land to sea.

Now think of the whole earth. In the hot tropical countries around the equator, enormous quantities of hot air rise up and spread out to both the north and south, away from the equator. This draws in cooler — but still quite warm — winds. At the icy poles, the air falls and there is a continuous flow of cold air away from the poles. Between the equatorial winds and the polar winds, in temperate zones, the winds blow towards the

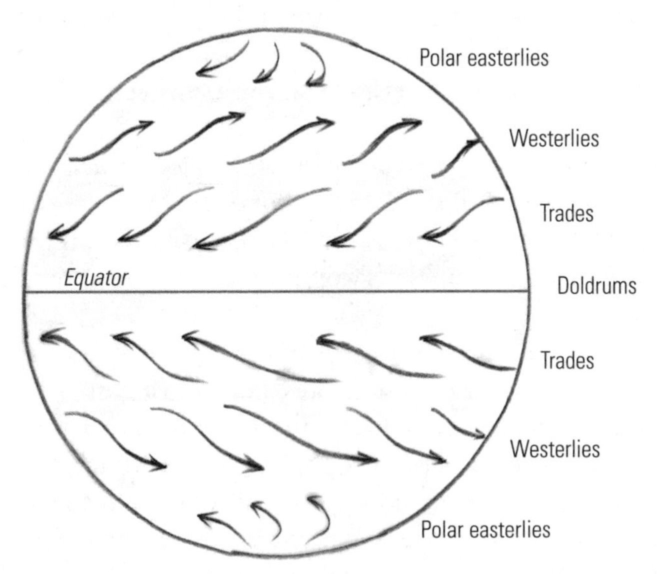

The winds of the world

poles. This circulation of air makes winds blow over the oceans all the year and, because the earth is always rotating, these winds do not just go north and south, they turn east and west too. On both sides of the equator there are the steady, reliable trade winds. Further north (and south) there are colder winds which blow the opposite way, from west to east, the westerlies. In the old days, sailing ships made use of these winds for their journeys. So the air around us is forever moving because warm air goes up, cold air sinks down and the earth turns.

But sometimes this pattern of hot air and cold air is not just a slow and steady circulation; it becomes instead a madly turning spinning-top. These mad spinning-tops of air happen in warm, tropical regions and are called *hurricanes*. Sometimes these hurricanes strike the coast. This happens quite often on the southern coast of the United States. Then trees are flattened like matches and houses are broken apart as if a bulldozer had gone through them. Cars and even houses can be blown a distance of half a mile.

Strangely, right in the centre of this spinning-top of air, there is no wind at all. This still centre of the storm called the *eye* of the hurricane. Around is a swirling hell of storm and water, but in the centre it is calm. You can even see the blue sky and the sun above.

Like all the winds of the earth the hurricanes too come from the process you see when smoke rises from a fire: hot air goes up, cool air sinks down. And what moves the air in the world? It is the warmth of the sun. Air as well as water, both are moved by the sun.

The way the sun warms the earth and warms the water has another effect. Think of summer and winter in Britain. This island is surrounded on all sides by the sea. The sea around our shores may look and feel very cold to you in winter; but because the sea cools down slowly it is still warmer than the land. The warm air rises from the sea and it spreads and comes to us. So we do not usually get very harsh winters with lots of ice and snow. But in summer the sea around us is still cool,

Floris
Books

compared with the land, and we get a good deal of cool air in summer. So we rarely have a really hot summer, and cool air brings the clouds and rain!

Now think of a country right in the middle of Europe — so far from the sea that no air can reach it directly from the sea. It will be very hot in the summer, and very cold in winter. So Britain's *climate* as it is called — the weather over the whole year — is different from the climate of, say, Austria for two reasons: because Britain firstly is further north and secondly is surrounded by sea.

13

Glaciers

We have looked at earth, water, air and also at the warmth of the sun; how it moves the winds and the waters of the world. All these things have shaped and formed the world in which we live and they go on shaping and forming it. But there is something else that has played a great part in shaping the world, and that is snow and ice.

On the peaks of high mountains like the Alps the snow does not melt; it stays from one winter to the next. If winter after winter more and more snow falls, over hundreds or thousands of years, the mountain peaks of the Alps or the Himalayas would by now have a covering of snow higher than the mountains themselves. But why do they not have such high caps of snow?

When there is such a thick, deep layer of snow, the snow at the bottom is under great pressure, for all the snow above is pressing down on it. When snow is pressed down hard, it changes first from fluffy snow to hard, compact ice, then it melts. When all the snow and ice lies on a slope and the ice at the bottom turns partly to melt-water, the whole thing very slowly slides over the rocks below. The whole ice-field creeps and slips slowly further and further downhill.

Such a moving ice-field is called a *glacier*. It is really an ice river, but an ice river that flows only very, very slowly downhill. How far can it flow downhill? Until it comes to the level where the air gets warm enough in summer to melt all ice and snow, and there the glacier stops. Streams run from it, but it does not all melt away for there is always more ice coming down from above and more snow falling in the heights. It is a terrific sight, a glacier, a vast field of ice — 40, 50, 100 metres thick in some places — crossed by cracks called *crevasses*, some quite narrow, but some very wide and deep. It is these cracks in the ice which make crossing a glacier difficult and dangerous. Looking down one of these gaps you see deeper layers of ice, and they appear blue.

As you look at it, the glacier seems to be as hard and still as rock — it does not seem to move at all — but it only seems so. If a flag is stuck into the ice, you will still be able to find it a few days later, several meters further down, and in time it will come to the end of the glacier. This giant ice field is not quick, it has a slow downward-creeping motion. (Some hardly move at all, but fast ones can move several metres each day).

The glacier is also strewn with boulders, rocks and stones which have been broken from the rocks by the river of ice. All these boulders and stones are slowly moved to the end, where the glacier melts, and then they come to lie on the bare ground. This "depositing," this bringing down of boulders and stones, has been going on for many years and so every glacier has a great fringe, a "hem" of rubble, boulders and stones. This great fringe at the glacier's end is called a *moraine.*

Now the strange thing is that there are many places in the

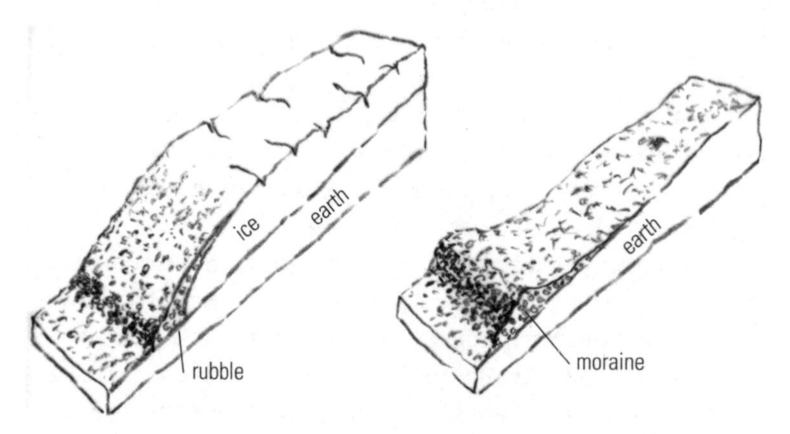

A glacier brings rubble and when it melts leaves it as a moraine

world where you can find quite enormous moraines, large fields of rubble, boulders and stones; but they are far away down in valleys and in plains, far from any glacier. For a long time no one could think of an explanation of how these moraines came to be there. In the end a Swiss scientist called Louis Agassiz found the explanation. In ancient times the glaciers must have reached much further down than they do now. That was the *Ice Age* and it ended only about 10 000 years ago. That's how we came to know that there was an ice age.

Now think of the first picture we described: earth and mountain. In that picture the earth as well as the mountain had an ice cap. During the Ice Age both ice caps reached much further. Most of Britain was covered by ice so deep that our mountains were all completely hidden beneath these terrible glaciers. Much of Europe and North America were also covered.

The underside of a glacier is just like an enormous rasp (think of woodwork lessons). As it flows downhill it rasps the rock below. This "rasp" is not made of ice alone; it has countless rocks and stones in it. As the glacier flows along these all grind and scrape the bedrock below. Even today there are many places where you can see the parallel scratches left on the surface of a rock by glaciers during the Ice Age. So our mountains have been very much worn down by the Ice Age glaciers. Between the smoothly rounded hills the ice rasp has hollowed furrows and these have become our smooth, U-shaped valleys. Often a stream now flows along the valley floor, but it was not the little stream which cut out such a big valley — it was a mighty glacier during the Ice Age. The rounded, "rolling" landscape of Britain has been shaped by glaciers.

If you rasp wood you get sawdust from all the wood rasped away; the glaciers, these ice rasps, have also left behind a vast amount of boulders, stones and rock dust. This mixture is called *boulder clay*. At the end of the Ice Age there were vast piles of boulder clay along with lakes ponds and rivers left by the melted ice. Quite quickly plants began to grow and animals moved in to live, so that the boulder clay became mixed with plant and animal remains. This all created wonderful rich topsoil. Some of the best farmland in the world, including the fertile land of the Lothians here, is based on the rock and clay left by the glaciers of the Ice Age.

Why did the ice spread? And why has it since then melted and retreated to the far north and the great heights? No one knows for certain but there are many possibilities. It may be due to changes in the atmosphere of the earth — more or less carbon dioxide in the air — a kind of "breathing" of the earth. It may be because of small changes in the strength of the sunlight falling on the earth. It may be because so many volcanoes erupt at once that their smoke and dust dim the light of the sun for a while. It may be because of changes in the winds or the way warm and cold water move around in the oceans. Or perhaps it is something quite different. We cannot be sure, but it is possible that the glaciers will creep down on us again in a few thousand years' time.

14

The Story of Arthur's Seat in Edinburgh

We have heard of the ways in which mountains are "born" and how, through water and wind they are worn down again. I now want to describe a hill here in Edinburgh, Arthur's Seat. Every mountain tells its own story; and the people who study the rocks and mountains, the geologists, can read the story of a mountain from the layers and rocks.

The story of Arthur's Seat began many millions of years ago, long before Arthur's Seat came into existence. Nearby, to the south of here, there were great volcanic eruptions. In time, the lava and ash from these eruptions formed mountains and we can still see their worn-down remains today: the Pentlands, the Braid Hills and Blackford Hill. But the weather, the climate here, was very different from what it is now. It was hot, so hot that this part of the world was a desert. It is strange to think that this was so, but that is the story that the rocks tell us. There was once a time when the lowlands of Scotland were a desert, a desert deeply covered with sand.

After many millions of years the land began to sink (or the sea rose) and the desert with its thick layer of sand became a sea-bed. The sand of the desert was pressed down and became sandstone. But then, as more millions of years passed, something else happened. The water above the sandstone became a quiet tropical sea, with tree-lined beaches and lagoons full of life: fish, shrimps, mussels.

Then, in this tropical sea a terrific volcanic eruption took place. It must have been an enormous eruption for the lava forced its way right through layers of stone and millions of tons of lava came pouring out: lava mixed with the rocks and stones that had been pushed out by the eruption. First a volcano appeared where the Castle Rock now stands and then, through this mighty eruption, the first shape of the volcano we call Arthur's Seat came about.

This was formed by ash and lava together with all the rock and stone that the lava had pushed out in the eruption. But the *pipe,* the shaft through which the lava had come, was free and more lava came in powerful eruptions. Lava and ash covered everything beneath it.

In time the volcano of Arthur's Seat which had been smoking for a long time, gave less and less smoke; it became dormant. The lava inside the shaft hardened and cooled and the volcano became extinct; it became a dead volcano. Then the land sank down and Arthur's Seat slowly disappeared beneath the waves and became buried under thick layers of sediment at the bottom of the sea. In this sea there were coral reefs and teeming marine life — beds of shelly limestone formed at the bottom of this sea. At the same time great nearby rivers flowed into the sea, spreading out and depositing huge amounts of sand; in time this formed thick beds of grey sandstone.

Then the sea-level went down, the land rose and the land that had been sea-bed became covered by a strange forest in which all kinds of creatures lived. This was the time when many coal-seams were slowly formed, one above the other. Millions of years later these coal seams would be mined and millions of tons of coal dug out for use in homes and industry. But, deep down, there is still a huge amount of this coal left — the last remains of those ancient forests which once grew all around here and over much of the world.

Then there was more volcanic activity and molten rock forced its way upwards. This time the glowing, fiery magma did not break through to the surface and it hardened, deep underground, into huge slabs of basalt. (Some of these came to the surface — where we can now see them — but only much later, because of deep movements in the earth's crust). Salisbury Crags date from this time. The Crags look reddish, not dark like basalt. But basalt contains iron, and over the years its surface "weathers" to the reddish colour which we see. The hard basalt hills have been quarried and the rock used for building and road making. The quarrymen call this rock *whin.* But whin is also a name for the tough *gorse* plant which likes to grow on the stony basalt soil. And every spring the gorse or whin sets these volcanic hills aglow once again with its bright covering of golden flowers.

More time passed: the land rose again and the sea receded. The crust of the earth began to fold. This powerful folding of the earth, brought about by the movement of continents, caused earthquakes — the rocks slowly bent, folded and slid into humps and troughs, domes and basins. In this upheaval Arthur's Seat (together with the younger Crags) was lifted again from beneath the sea while the rocks all around were shifted, bent and mixed up. Many of the higher up layers of rocks were worn away, while others — like the Midlothian Coalfield — were preserved. The softer rocks which had covered over Arthur's Seat were worn away, but the hard volcanic rock was not, and so Arthur's Seat stood high above the land once more.

Then, in more "recent" times, the Ice Age came and ice covered the whole land. The deep ice-sheet flowed slowly towards the east and ground away at all the rocks, including Arthur's Seat and the Crags, but even the power of the ice could not remove them. Finally the Ice Age went away and by this time, which was "only" about 10 000 years ago, Arthur's Seat looked more or less as it looks today.

So this is the story of Arthur's Seat. And now think of all the stages this mountain has seen. It has seen tropical heat and it has been deep under glaciers in the ice age. It has been dry land, and it has been a sea-bed, deep under the waves — and then dry land again. It has seen the slow process of mountain-folding as well as the sudden terrific bursts of volcanic eruptions. So Arthur's Seat is a mountain that has seen all the things we have talked about.

Astronomy

15

The Heart and the Sun

If we think about what the most important thing is that a human being needs, we would probably say food. But if we think it over, we will realize that water is even more important. It is easier to go hungry than to go thirsty, and people lost in the desert can last many days without food, if they have water to drink. But a human being can go perhaps two days without water. But how long could anybody go without air? Not even two minutes. It is fortunate that we don't have to work or to pay for air, that it is always there for us to breathe. We must breathe to live.

The rate, or speed of breathing changes. We breathe faster when we run and slower when we sit. If we take our normal breathing and we count breathing in and breathing out as one breath, the average number of breaths per minute is 18. As there are 60 minutes to the hour, we breathe 18 x 60, that is 1 080 times per hour. And in 24 hours that makes 1 080 x 24, or 25 920 breaths every day. That's an average of course, if we swim, run or climb during the day, it will be a good deal more.

When we run or climb hard, not only does our breath come faster, but our heart beats much faster. It is really our heart beating faster that makes our lungs need more air. How slow or how fast we breathe depends on our heart. That we breathe normally about 18 times a minute depends on the heart. For every four heartbeats there is one breath, and if the heart beats quickly then the breath has to come quickly. So heartbeat and breath go in a steady rhythm together, day and night.

There is another rhythm, also a kind of breathing, but a very slow breathing. It takes much longer, but we need it just as much as we need air. It is the rhythm of sleeping and waking. When we fall asleep it is as if something were going away from us, like slowly breathing something out. And when we awake, it comes back; we are breathing in again. The rhythm of sleeping and waking is also a kind of breathing.

The real breathing, taking in air and letting it out again, is ruled by the heart. Of course, we can breathe quickly or slowly, just for fun; but if we don't interfere, the heart rules and controls the breathing. The heart is the organ which knows how many breaths the body needs. This is because the blood goes through the heart, and the heart knows from the blood how much air the body needs.

The other kind of breathing, the "sleeping and waking," is certainly not ruled by the heart. The heart goes on in its own rhythm, day and night. Sleeping and waking are for us modern people not what they were in ancient times when people went to sleep as darkness fell and woke up at daybreak. They rose with the sun, and they went to sleep when the sun set. Their sleeping and waking was ruled by the sun. Even today, though we stay up long after it gets dark, we feel it is healthier, more wholesome, to sleep at night and be awake in the daytime than the other way round. We still feel it is better to be awake with the sun.

So the shorter rhythm of breathing in and out is ruled by the heart, and the longer kind of breathing — sleeping and waking — is ruled by the sun, that is like a great heart in the world outside. Though the sun is a kind ruler who leaves us modern people a good deal of freedom.

There is a third kind of "breathing." This time it is not a human being who breathes in and out. It is something much larger and so the breathing is very slow. Again it is not a breathing of air, but of something quite different, yet it is still a rhythm.

In spring all the green leaves and flowers appear, then more and more flowers appear towards the summer, and then autumn comes and fields, gardens and trees become bare. This too is a kind of breathing: it is as if the earth breathes out in spring and summer and breathes in again in autumn and winter. Only the earth is so vast that it takes a whole year for such a single breathing in and out. But this long, slow breathing of the earth is ruled by the sun. The rhythm of the seasons is ruled by the sun.

The "breathing" of the earth is not only so much slower than ours, it is altogether different. For when it is winter in Britain and Europe, it is also winter in North America and Russia. But on the other side of the globe — in Australia, in Southern Africa, in South America — it is summer, it is the warm season. So while one half of the earth is breathing in the other half is breathing out.

Things are even more complicated. For if we went north to the Arctic, to the North Pole we would find it frozen all the time; it is winter during the whole year. And if we went to the equator — to Africa or Central America — there are again no seasons; it is summer all year.

It is as if the earth would always keep a balance. When it is summer in one half it is winter in the other half. If it is spring here it must be autumn there. And if it is winter all the time at the poles, it is summer all the time on the equator.

All this is ruled by the sun. As the heart rules the rhythm of our breathing, so the sun rules the whole complicated rhythm of summer and winter of the earth.

16

The Sun's Daily Movement

Let us look at the connection between the sun and the breathing of the earth. There is permanent winter at the poles, and permanent summer on the equator, while in between the seasons change.

Let us imagine that we are on a great wide plain, so that the horizon forms a great circle. The path of the sun will go as a semicircle across the circle of the horizon, rising at one point and setting at the other.

This diagram showing the circle of the horizon and the semicircle of the sun is correct only for one particular part of the earth, for the equator. If we lived on the equator, we would see the sun rise at an angle of 90°, at a right angle to the horizon. And it also sets at a right angle to it. At noon the sun stands much higher in the sky than it stands here. It stands so high that it shines down into the deepest well. It stands so high that things throw no shadow at noon — your shadow disappears

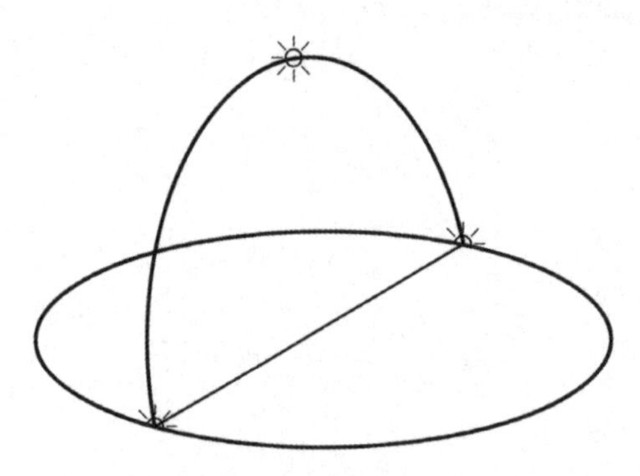

The sun's path at the equator

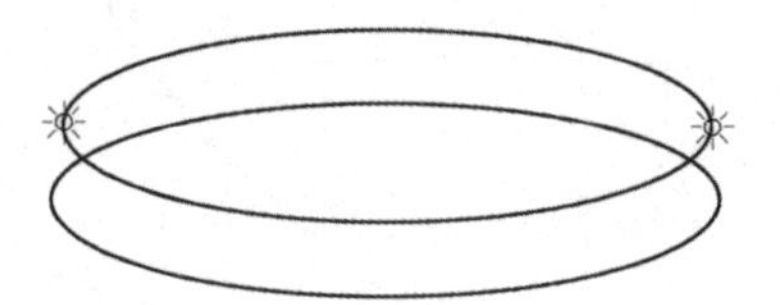

The sun's path at the North Pole

under your feet. The point exactly overhead, where the sun stands at noon at the equator, is called the *zenith.*

On the equator day and night always have the same length; there are no shorter days or longer days. And there is also no difference between the seasons. It does not get colder some months and warmer in other months, it is always hot. Only the rains make a difference, but this has nothing to do with the path of the sun. One could say at the equator there is no year with its seasons, there is only the day.

At the North Pole or the South Pole things are quite different. In summer in the course of 24 hours the sun describes a full circle above the horizon, and so there is no night. But this circle of the sun is not very high, at its highest it is 23½° above the horizon, only a little higher than the sun at noon in Britain in December.

This circle of the sun's path in the sky is really a spiral, towards autumn the sun circles at ever lower heights but still remaining above the horizon for 24 hours. Towards the end of September it reaches the horizon, slowly setting. Then for six months during winter it stays below the horizon. For months it is night and only a faint glow can be seen on the horizon.

Comparing equator and pole, we see that at the equator you can tell the time of the day by looking at the sun, but not the time of the year. It is as if there is no year, only the day. At the pole you cannot tell the time of the day because the sun is always at the same height. But you can easily tell the time of the year. At the poles there is no day, only the year.

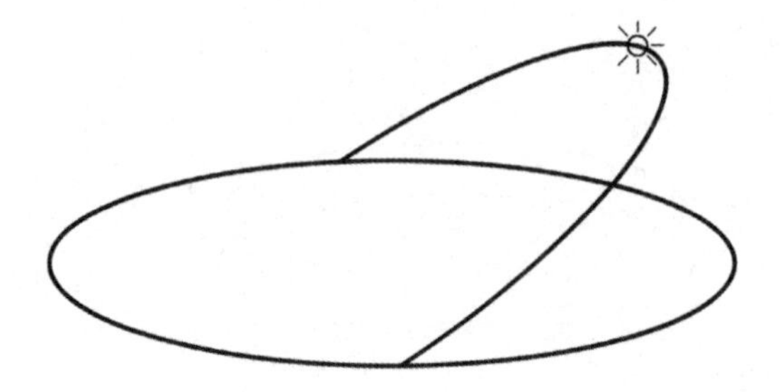

The sun's path in temperate latitudes

While at the equator the sun rises at right angles to the horizon, and at the pole the sun moves in circles parallel to the horizon, in our part of the world the sun rises at an angle to the horizon. We can show this in a diagram of the path of the sun in our part of the world.

At the equator at noon the sun appears overhead, in the zenith. In temperate zones (like Britain) it is somewhere between the horizon and overhead at noon. And at the pole it is close to the horizon. But this shows us that it is not really the sun that is higher or lower, it is we, the human beings who look at the sun from different angles. We can see how this comes about from the diagram of the globe of the earth. The person on the equator sees the sun straight overhead, in the zenith, the person in our part of the world sees it lower in the sky, and the person at the pole will see it very low on the horizon.

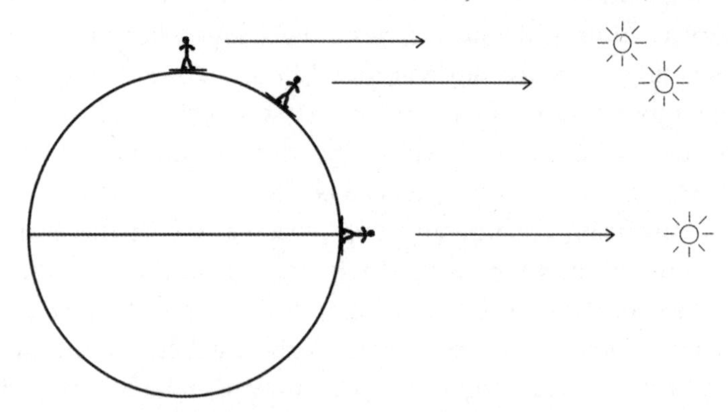

The height of the noon sun at different latitudes

The angle at which we see the sun at noon tells us where we are on earth between equator and pole. The nearer we are to the equator, the higher the sun will be at noon, and the closer we are to the pole, the lower the sun will be. So we know that in different parts of the earth, in different *latitudes* as it is called, the sun appears at different angles at noon.

But this angle at which the light of the sun comes to us makes all the difference to the whole of nature where we live. We can see it from this: If we draw a short line, and then draw as many sun-rays as we can, parallel to each other and falling at right angles on this line. And as we draw them, we count them. Now if we draw a line of the same length, but with parallel sun-rays falling on it from an acute angle, and we count again as we draw as many as we can; we'll find that there are far fewer rays at the acute angle than at the right angle.

Now we can understand that at a low angle the sun has less power, gives less light and less warmth than when it shines down at right angles. We can understand now why it is cold at the poles and hot on the equator and *temperate,* as it is called, in our part of the world.

How people dress, what kind of house they build, what kind of plants and animals live around them, all depends on the angle of the sunlight.

We can also observe the difference which the angles make in a single day. Even on a very hot summer's day, it is cooler in the early morning and towards evening — because the sun rays come in at a lower (or more acute) angle.

17

The Sun During the Year

We saw how the path of the sun appears differently in different parts of the earth. At the equator, where the path of the sun rises at a right angle to the horizon, there are no seasons of the year. At the poles, where the sun circles in the sky parallel to the horizon, there are no times of the day. And in our part of the world we have both times of the day and the seasons of the year. Now we shall see how these two are connected with the path of the sun.

In late autumn or early winter we have a chance to get up while it is still dark and see the sun rise (if it is not clouded). If we do this for several days and make a note of a landmark (like a house, tree or hill) where we see the sunrise. After a few days we shall see that the sunrise moves. And if we do the same in the evening we shall see that the sunset also moves. If each time we make a little drawing of what we see, then we would find that during autumn we find that sunrise moves towards the right and sunset moves left.*

If you hold your left arm in the direction of sunrise and your right arm in the direction where the sun was setting on one day, and you do the same a few days later, you will see that the angle between your arms becomes smaller, narrower. And if you watch the sun at 12 o'clock (when it is at its highest), you will see that every day it is a little lower in the sky.

* The movement is not so pronounced in lower latitudes, and also less pronounced towards the winter (and summer) solstice. Kovacs found one of the few examples of astronomical advantage of Edinburgh's relatively high latitude (56°N).

Sunrise creeping southward (to the right looking east) in late autumn

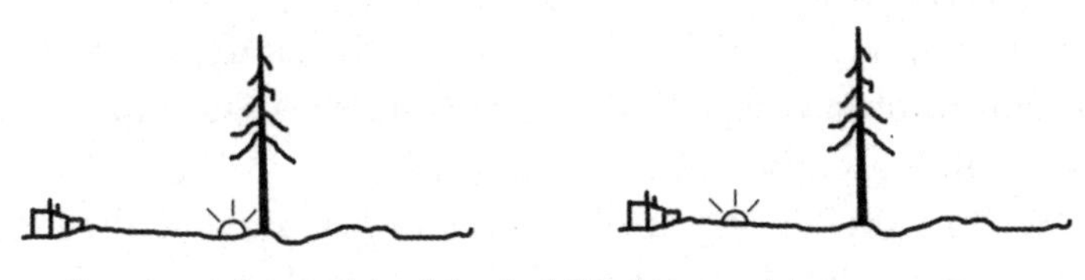

Sunset creeping southward (to the left looking west) in late autumn

As your arms between the directions of sunrise and sunset come closer together, and as your gaze towards the midday sun sinks lower, the days get shorter and shorter. From the movement of the arms and head you can see this is a kind of inbreathing.

So we are losing sunlight in two ways: the days get shorter and the angle of the midday sun also gets less. We receive less light and less warmth.

The shortening of the day and sinking of the midday sun will go on until December 21 or 22, the shortest day of the year. From December 23 onwards the angle of our arms between sunset and sunrise will widen, our eyes have to look higher and higher towards the midday sun. The days get longer. Then on March 20 or 21 day and night are of the same length, twelve hours.

During the time from December 23 onwards the points of sunset and sunrise move, but not as they did before, now sunrise moves to the left and the sunset to the right. And they continue this journey, the sun being higher at midday and its journey over the sky lasting longer, until June 21, the longest day of the year. The movement of our arms and head following the sunrise, sunset and noon become like an outbreathing.

From June 21 the days shorten again, the midday sun sinks lower and on September 22 or 23 day and night are of equal length again.

So we have one longest and one shortest day in the year, and two days when day and night are equally long, one in spring and one in autumn. These days are called *equinoxes* (equal night). March 21 is the spring equinox, and September 23 is the autumn equinox. The longest and the shortest day, when the sun turns, are called the *solstice.* June 22 is the summer solstice, and December 22 the winter solstice.

We can now see the height of the midday sun and the length of the day go together. The higher the sun, the more light rays, the more warmth.

Let us look once more at the movements of sunset and sunrise. When we first looked at this we found in late autumn sunrise moves to the right and sunset to left. If we think of the cardinal points of the compass, to the right of east is south, and to the left of west is south. That means both points move south. It seems strange, but in winter both sunset and sunrise move south. At the winter solstice sunrise is in the south-east, sunset is in the south-west. And in the summer sunrise and sunset both move north. On the longest day, at the summer solstice, sunset and sunrise are furthest north, in the north-east and the north-west.* Only on the two equinox days, March 21 and September 23, does the sun rise exactly in the east and set exactly in the west.

During the course of one day the sun rises and sets, it goes up and goes down. And during the course of a year the path of the sun goes up and goes down. Additionally, the path of the sun moves northwards for half a year (from the winter to the summer solstice) and then southwards for half a year (from the summer to the winter solstice).

This twofold movement of the path of the sun (up–down,

* In lower latitudes solstice sunrise and sunset will not be as far south or north as described above. This happens to hold true fairly exactly for Edinburgh's latitude. Sunrise and sunset at the equinox are always east and west everywhere in the world.

north–south) can best be seen in the polar regions. If we were standing at the North Pole any direction in which we look or walk is south. The North Pole is the furthest we can go north on earth, and any step away from the North Pole is a step south.

In summer at the North Pole you see the sun circling in the sky. Towards autumn it sinks lower, and in which direction does the path of the sun go as it sinks? It moves to the south. In the long winter nights when the sun does not rise at all in the Arctic the sun is too far south to be seen.

On the equator this twofold movement of the sun's path can also be seen but not as clearly as at the Pole. In December the sun rises and sets a little to the south of east and west, and in June it rises and sets a little to the north. There is no difference in the length of the day and only a little variation of the height of the sun at noon. In December it is a little to the south of the zenith, the point overhead, and in June a little to the north. Only at the equinoxes does the noon sun go exactly overhead, through the zenith, at the equator.

The sun's path is not really a circle but, during the course of the year, it is a spiral. For the sun does not return to the point where it was the day before. At the North Pole we would be able to see the sun's path spiralling up from spring equinox to summer solstice and then spiralling down until the autumn equinox. We would be able to see the entire spiral for half a year.

At the equator we would be able to see half the spiral for the whole year (as the sun is above the horizon for twelve hours every day). In our part of the world we have something in between, for half the year (the summer half) we can see more than half of the spiral, and in the winter half of the year we see less than half of the spiral.

This spiral movement that goes up and down gives us the seasons of the year which are such a familiar part of our life.

18

The Calendar

As long as people only hunted, as in the early Stone Age, they did not have to watch the seasons very carefully. They could hunt animals in winter as well as in summer. But when agriculture began, in the New Stone Age, in ancient Persian times, people began to sow and harvest crops and the change of seasons became very important.

The first farmers, the first peasants, were also the first people who needed a calendar. Different crops — grain and vegetables — have to be sown and harvested at different times of the year, so the farmers had to know what time of the year it was. But there were no calendars (there were no books at all) and the people had to find ways to ensure they knew when to sow their crops.

In Britain, for instance, the wise priests who guided the people built the great stone circle of Stonehenge. It was a holy place, but it was also a kind of enormous calendar. The stones were landmarks for the path of the sun through the year. The priests watched specially the shadows cast by the stones, for the shadows change in the course of the year. The shadows were like the hands of a great clock which showed not hours, but months and from the shadows the priests could tell the people when it was time to sow or to shear the sheep.

Far away in Asia, in Babylonia, the priests also observed the path of the sun. Babylonia was already a great civilization with mighty cities, and the priests built great observatories from where they carefully and patiently watched the path of sun, moon and stars.

It was not only for agriculture, for the priests of Babylonia said that just as certain plants grow only at a certain time of the year, so a great warrior, for instance, can only be born in a certain time of the year, in the month after the spring equinox. But a boy born, say, in February would make a learned priest. It was like this for every month, even for every day of the year. When a child was born, the priests of Babylonia told the parents what kind of destiny this child would have, according to the time of birth.

That is why the Babylonian priests observed sun, moon and stars so carefully and why they became the people who gave us the division of the year into twelve months and into fifty-two weeks and the week into seven days, and the day into twice twelve hours. All these measurements of time go back to the priests of Babylonia, the land between the Euphrates and the Tigris.

To find how many days were in the year, the priests of Babylonia measured the shadow of a pillar or obelisk very exactly at noon on the longest day of the year, the summer solstice. At that moment the shadow would be shorter than at any other time of the year. And then they counted the days until the shadow was again of the same short length. They counted 365 days. But at noon on the 365th day the shadow was a little longer which meant the sun was not where it had been the last time. So they waited another 365 days, and after that time the shadow was still longer than the first time. And the same thing happened after another 365 days. So they had gone three times through 365 days without getting the same length of shadow on the longest day of the year. The next time, the fourth time, they *did* get the same shadow lengths as the first time, but after 366 days, not 365 days. The reason was that the real length of the year is not 365 days, but three hundred and sixty-five and a quarter days, or 365 days and 6 hours. And that is why we have one extra day — February 29 — every four years, the *leap year.*

The priests of Babylonia discovered by measuring shadows that the true length of the year is 365¼ days. The Romans found this out by another way. They at first counted 365 days, year after year. After four years they were only one day ahead of the sun and no one noticed any difference. But after one hundred and twenty years they were one month ahead of the sun — they called the month April, but the sun was still at the equinox in March. Different things were tried to make things right again, but it became worse until the time of Julius Caesar. Caesar had travelled in Egypt where the priests knew that the year was longer than 365 days. It was Julius Caesar who changed the Roman calendar and introduced the leap year. And the calendar he introduced is called the Julian calendar.

But the length of the year is not exactly 365 days and 6 hours, it is 365 days 5 hours 48 minutes and 45 seconds. Julius Caesar did not know this, but this little mistake made his calendar wrong. Every year people were eleven and a quarter minutes behind the sun. That's very little in one year, but after 1600 years this little difference had become ten days. People were ten days behind the sun, or you could say the sun was ten days ahead of the calendar. If this had gone on people would have celebrated Christmas a month after the winter solstice instead of two days after it.

In 1582 the Pope of that time, Gregory XIII, made another change in the calendar. First, to get things back on track, ten days were struck out. Thursday, October 4, was followed not by the 5th, but by Friday, October 15. Many people did not like that at all. In Britain (where this change was made seventy years later because of a quarrel with the Pope) there were riots because people thought the government had robbed them of ten days of their lives.

The ten days struck off the calendar only corrected the past mistake. In future to adjust the year for the eleven and a quarter minutes difference, the rule that is still followed to this day, is that:

— every fourth year is a leap year (that is any year that can be divided by four, like 1868, 1992 or 2016)
— but every century is *not* a leap year, *except* those which can be divided by 400 (so 1700, 1800, 1900 were not leap years, but 1600 and 2000 were leap years).

The year seems quite a simple thing, but we now see what a complicated business it is. And even this complicated rule, called the Gregorian calendar after the Pope, is still not quite correct. We are now gaining 11 seconds every year, but as it will take about 4000 years for this error to grow to a full day, we don't need to worry about it.

At the root of the matter is the problem that the real length of a year cannot be divided by any whole number without a remainder, or if calculated in decimals it would go on endlessly. The length of a year is an irrational number, as such numbers are called. The path of the sun cannot be put into simple numbers.

19

Sundials and Time

We heard how the wise men in Babylonia watched the shadow of a pillar and in ancient Britain watched the shadow of the stones at Stonehenge. From the shadow they could tell the people when it was time to plant grain and when it was time to shear the sheep.

The shadow cast by the sun is altogether very interesting. At the poles — where the sun stays at the same height all day as it goes round the sky — what would the shadow of a stick put straight into the snow, show? The shadow would stay the same length all day; it would not get shorter or longer, as the sun remains at the same height. And it would wander around the stick in an exact circle. (To be quite precise it would moves in a spiral, getting a little shorter or longer as it went round, depending on whether the sun was climbing up or down.) We know that the higher the sun is in the sky the shorter the shadow will be. At the poles the shadow will be quite long for the sun does not get very high.

We can also easily imagine the shadow of a stick at the equator. At one of the equinoxes the sun rises exactly in the east, goes right up through the zenith (exactly overhead) and then sets exactly in the West. At sunrise and at sunset the shadow will be very long, and at noon when the sun is in the zenith, there will be no shadow at all. When the sun rises exactly in the east the shadow will point straight west, and at sunset it will point straight east. A line from true east to true west is a straight line, so on that day the shadow of the stick will not move *round* the stick at all, it will remain a straight line, first pointing west, then getting shorter and shorter until it disappears under the stick, and then pointing east getting longer and longer, but all the time it will keep on the same straight line.

So even in the shadows the pole and the equator are opposites. In the course of a day at the pole the shadow keeps the same length, and goes round in a circle, while on the equator the shadow changes its length, from long to nothing to long, and moves in a straight line.

For us who live in temperate climates the shadow will not move in a straight line or in a circle. It will change its length, longer in the morning and evening and shortest at noon, but what shape will the path of the shadow make if it's neither a circle nor a straight line?

We can find the answer by experiment. Take a sheet of paper and fix it to a board and fix a large needle or nail vertically to the board. Now put the board on a window sill or flat surface outside, making sure it is horizontal, and that it catches the sun for as much of the day as possible. Then every half hour, or as often as possible, mark the point where the shadow ends.

If all is set up well, and we have good weather, we shall have a complete set of points. When we connect these points we find they make a curve. If we have done this in winter when the sun rises in the south-east there will be a long shadow pointing north-west, at noon when the sun is in the south there will be a shorter shadow to the north, and at sunset (in the south-west) there will be a long shadow pointing to the north-east.

If we did this every month for a year we would find a set of curves as shown overleaf. Among these curves is a straight line, which is the shadow cast by the sun at the equinox when it rises exactly in the east, and sets exactly in the west. But as we are not on the equator the top of the shadow does not disappear exactly under the stick, but still has a length, and so is north of the stick.

A vertical stick at the North Pole would be a perfect *sundial.* Draw a circle round the stick and mark twenty-four hours

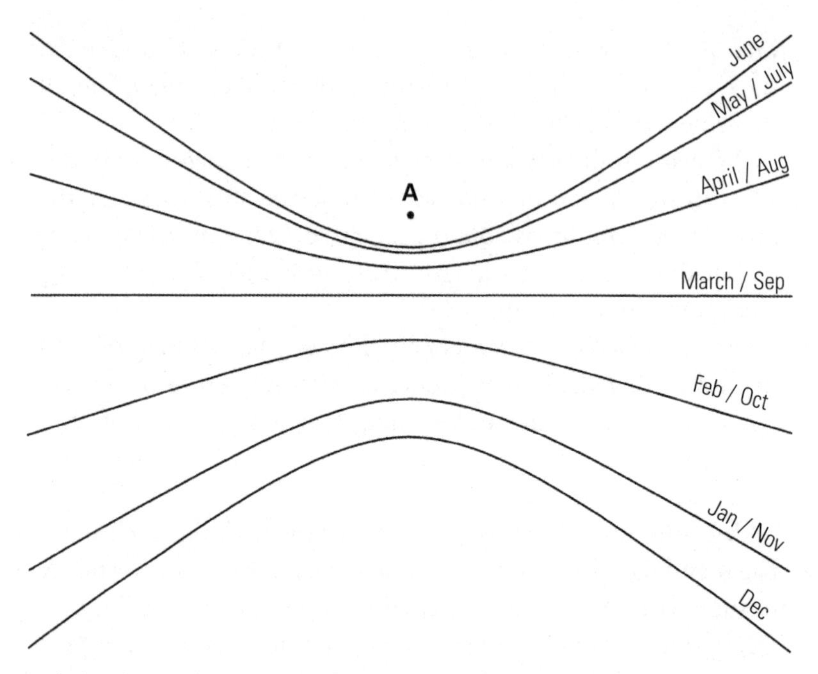

Shadow lines drawn once a month at the end point of the shadow of a rod (point A) *(after Baravalle)*

around it. The shadow would then show the time. But in our part of the world we would have to put the stick parallel to that stick at the Pole — for Edinburgh that would be at an angle of 56° to the horizontal (the latitude of Edinburgh). The *gnomon,* as the stick of a sundial called, is always at an angle. If you measured the angle you would find it is always the same as the latitude of the place where the sundial stands.

When the shadow is shortest it is noon our watch will not show exactly twelve o'clock, because our clocks do not go by the true time shown by the sun. During the course of the year, the sun sometimes moves a little faster, sometimes a little slower, while our clocks rigidly keep to the same speed. So there can be differences of up to 15 minutes between the sun's noon and clock noon.

Another thing which can make an even bigger difference is the following. The further west you go, the later the sun rises and sets, and the later it will be at its highest at noon. At the latitude of London (51½°) for every 17 km (11 miles) you go west, the sun rises one minute later. So Bristol's real or local time is ten minutes behind London's. Before the days of railways each town and village kept its own local time, and that worked fine for everyday life. But when the railways were built and trains ran to timetables, a standard time was introduced across the whole country. In Britain we keep to Greenwich Mean Time, the time of London's observatory in Greenwich.

Going further west, in New York it is noon five hours after it is noon in London. So it is still seven in the morning when it's noon in Britain. And going east, Beijing is eight hours ahead of Britain. While each country could have its own time, most countries in the world now keep to a time that is a whole number of hours different from Greenwich Mean Time. So travelling between Britain and Germany, for instance, there is exactly one hour time difference, or travelling between Sydney in Australia and Britain there is exactly 10 hours difference. Some countries stretch so far east and west that they have several time zones in the same country: the United States (including Alaska and Hawaii) has 6 time zones, each an hour apart, and Russia has 10!

20

The Circling Stars and the Pole Star

We have looked at the sun in connection with time: the time of day by the rising and setting of the sun itself, and the time of the year by the change in the whole path of the sun. And we have looked at the first clocks, the sundials, that tell the time of day, and the stones of Stonehenge which were a kind of clock for the year. Now we shall look at the directions of space, something that is also connected to the sun.

If, for instance on our holidays, we come to a town or village where we have never been before it takes some time until we know the layout of the streets. But a bee flying from its hive in search of nectar has no streets to go through, nor can bees see very far. Yet they often collect their nectar far out of sight of their hive, and always find their way back. They can even tell other bees where they have found a rich supply of nectar. Scientists who study bees discovered that the bees find their way by knowing at any time at which angle to the sun they have been flying. Strangely the movement of the sun over the sky does not matter to the bees, nor does it matter if the sky is overcast. They do not find their way by looking for landmarks on the ground (as we would do), they are guided by the position of the sun, and that it moves does not matter to them.

And scientists now believe that not only bees but also birds which fly south in winter (like swallows which fly all the way to Africa) use the sun to guide them in the right direction.

Birds and bees are born with this wisdom — or *instinct* as

it is called — which tells them how to guide themselves with the help of the sun. We human beings have not such instincts; we look far landmarks — the streets of a city, roads and paths in the countryside, or hills, trees and rivers when there are no roads.

But there are no landmarks on the sea. And when the first people took to boats they also had to look up to the sky, to the sun, to give them the direction in which they were going.

When in Greek or Roman times a ship sailed from Italy to Phoenicia — that is to Asia — they sailed in the direction of the rising sun. And that is why they called these lands of Asia the Orient. The word "Orient" means "rising." The Orient is the land which lies where we see the sun rising. And we have the verb "orientate," which means to find one's way about. "I have to orientate myself," means "I have to find out if I am going in the right direction." The phrase "to orientate oneself" goes right back to the times when sailors sailed by the sun.

When the same ship sailed back to Italy, to Europe, it sailed in the direction of the setting sun. And Europe became the Occident, meaning "setting," the land of the setting sun. Even in our time, in geography, we still speak of the countries of the Near East and the Far East; and the civilization of Europe and America is called "western" civilization. So we still use the sun to orientate ourselves in geography.

Most of our maps are made so that north is up, and south down, east on the right and west on the left. These four directions (the *cardinal points*) are taken from the sun's path over the sky.

All this goes back to the first sailors who had no landmarks and had to guide themselves by the sun. It is wonderful, when you come to think of it, that in order to find their way here on earth, the sailors of ancient times had to look up to the sun.

But none of the sailors of ancient times were ever far from land. Their journeys were short hops from one island to the next, and the Mediterranean Sea has so many islands that it was fairly easy to get across in any direction. Only when a storm blew a ship off course did the captain no longer know where he was with his boat, and had to trust his luck to find some land, a landmark, soon. None of these sailors of ancient times would have ventured out into seas that were unknown. If a ship did make a long journey, say from Gibraltar to Constantinople the sailors kept close to the African coast. And if they sailed from Gibraltar to Britain, they were never far from the coast of Portugal or France. Being close to the coast was the only way to know where they were. In winter when storms were more likely, ships did not sail at all.

The first European sailors who dared to venture out into unknown seas were the Vikings who sailed to Iceland, Greenland and Vinland (Flaki used ravens to give him the direction to the nearest land). They did this all without maps or compass, and they could never know exactly where they were on the sea; they could only say, "It took two days rowing north-west," or "three days rowing east." They did not have watches or clocks to tell them the exact time it had taken.

But the Greeks, the Romans and the Vikings all had to sail at night as well as by day, for they could not always find a convenient island or a harbour when darkness fell. When they looked up to the countless stars in the sky they saw that the whole sky with all the stars moved. Like the sun, the starry sky moved from east to west. The stars moved in circles, but all the circles of the stars had the same centre: the stars moved in concentric circles. In the centre of all the stars' circles there was one star that never moved at all, it always stayed in the same place. They called this star the Pole Star, or Polaris.

It can easily be found by looking at a group of bright stars called the Plough or the Great Bear. If you follow a line from the last two stars of the Plough you come to the Pole Star. If you watch it for a time you can see the whole Plough slowly swinging round the Pole Star, but the last two stars always point to it.

Not only is the Pole Star the only star that stands still, but it also stands exactly in the north. If you look in the direction

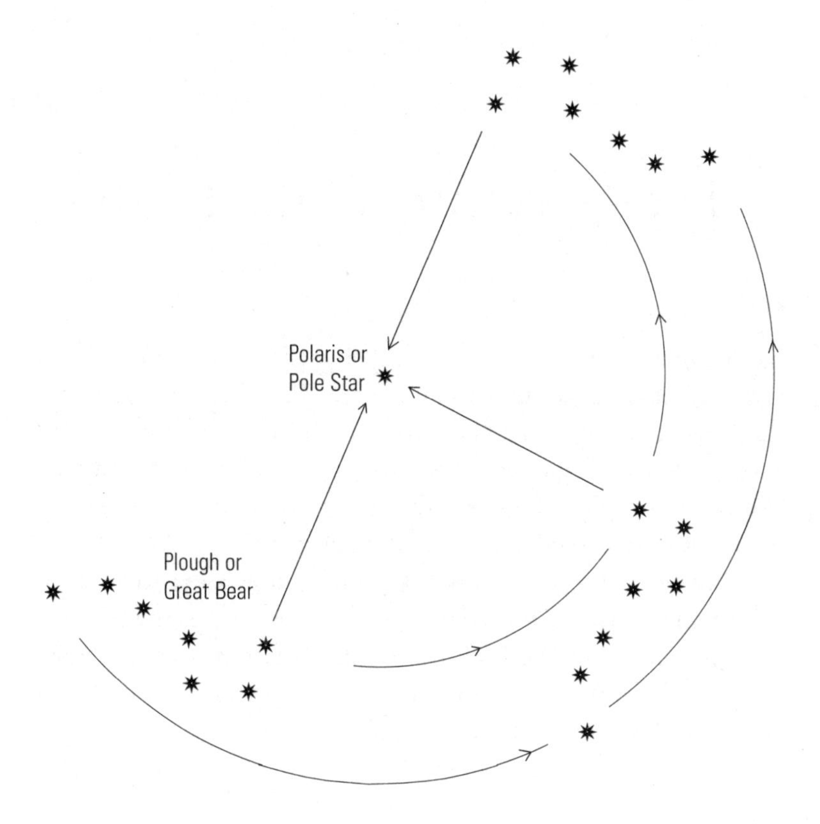

The Plough circling the Pole Star, the last two stars always pointing north

of the Pole Star you are looking exactly north. We can see why the Pole Star was so important to sailors; they could always find it on a cloudless night and it showed them exactly where north was. No matter whether they sailed east, south or north-west, they could get their bearings by saying: if the Pole Star is there in the north, then east is at a right angle, and so on.

Sailors in ancient times could orientate themselves in the day time by the sun which is in the south at midday, and at night time by the Pole Star which is always in the north.

21

The Curvature of the Earth

Sailors in ancient time could not go by landmarks on the open seas: there are no landmarks on the sea. They had to look up to the heavens to orientate themselves, to find their way here on earth. They were guided by the sun in the daytime and by the Pole Star at night. However, there were two things the sailors of ancient Greece noticed which puzzled them.

One puzzling thing was that when they sailed towards a mountainous island they first saw only the tops of the mountains peeping over the horizon. As they came nearer they saw more of the mountains, and only when they came still closer could they see the shoreline of the island itself.

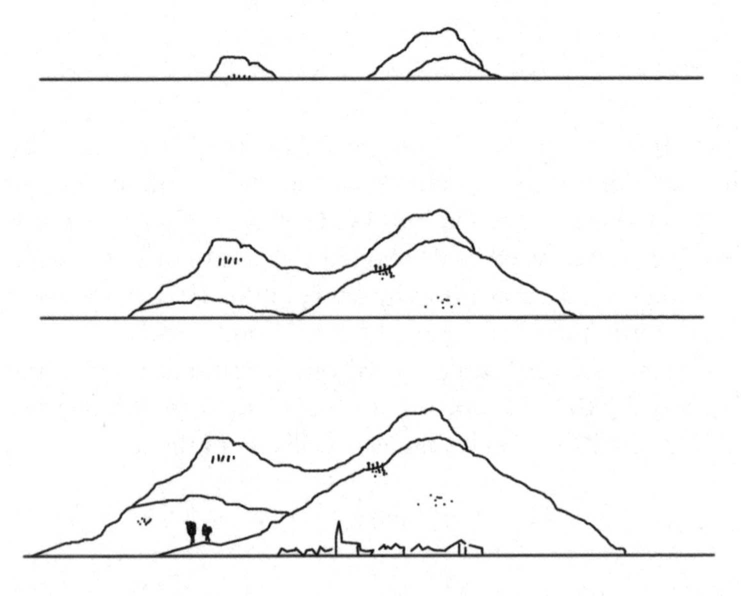

It was rather like when we walk up a hill and on the other side is a village with a high church tower. At first we will only see the top of the tower over the hilltop, then we will see more of the tower, and only when we get to the top will we see the whole village. And the sailors of ancient times, looking at the flat expanse of sea before them, wondered why it was that this flat, calm sea (on a calm day) was like a hill when they approached an island. Yet they knew they were not sailing uphill!

The other puzzling thing was the Pole Star itself. In the Mediterranean, where most of the sailing was done, the Pole Star is not very high over the horizon. If you held one arm horizontally and lifted the other to point at the Pole Star, the angle between your arms would be no more than 35°. But when these sailors from the Mediterranean sailed north, say to the coast of Britain, the Pole Star stood higher in the sky; the angle was 50° or more above the horizon. They knew the Pole Star does not move, but the further north they went the higher the Pole Star stood, and further south it stood lower.

These were the two things which puzzled the Greek sailors: approaching an island over a flat sea the island appeared as if they were walking up a hill; and that the Pole Star stood higher in the north than in the south.

In Alexandria, the great city of learning and knowledge that Alexander the Great had founded in Egypt, there were wise and learned men who found the answer to the sailors' puzzles. These wise men said that the earth is round, it is a globe.

If the earth is round the look-out on a ship will see first the peaks of a mountain it will be like going up a hill. And if the

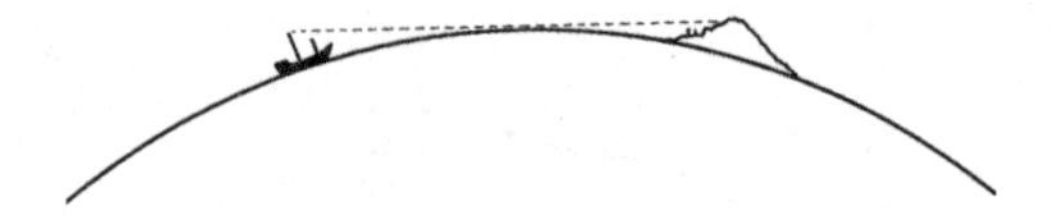

earth is round and the Pole Star stands in the zenith over the North Pole, then it will appear lower and lower the further you go south. And if you went as far as the equator, the Pole Star would be on the horizon. In the southern half of the world, for instance in Australia, you can't see the Pole Star at all. The stars there also go in a circle, but there is no visible star at the centre of the circles.

It is through these two observations made by sailors that the wise Greeks of Alexandria worked out that the earth is not flat but a round globe.

So why don't we normally notice that the earth is round? The larger a circle is, the less it curves or the flatter it is. The radius of the earth is nearly 4000 miles, and therefore the curve is so slight that we don't notice it.

Let us compare the Pole Star with the sun. At the North Pole the Pole star stands highest, in the zenith right above us. But the sun is very low and in winter disappears below the horizon. On the equator the sun stands highest at noon in the zenith, but the Pole star is on the horizon, and you cannot see it as the air on the horizon is never clear enough. The Pole Star is highest at the North Pole and lowest on the equator; the sun is highest on the equator and lowest at the North Pole. The Pole star stands still; the sun moves in a circle over the sky.

So the wise Greeks of Alexandria who never went very far north or very far south, just by thinking about the observations of sailors, found that the earth was round. It was by observing the sky — the sun and the stars — that they discovered something important about the earth. It was the sun and the stars, the Pole Star especially, which revealed the shape of the earth.

22

Longitude and Latitude

The wise Greeks of Alexandria came to the conclusion that the earth was not flat, but a globe. When the Roman Empire fell to the Barbarians, like so much other knowledge, this knowledge of the shape of the earth was forgotten and lost. It took a thousand years until people in Europe discovered anew what the Greeks had already known, that the earth is a globe.

If we take a ball and let light fall on it from one side, we see the shadow line: one half is in darkness, the other half is in the light, and the shadow line (which is not very sharp) divides the light part from the dark part. If this ball were the earth, the shadow line would go from the North Pole to the South Pole. And on one side would be night and on the other it would be day. All people along this shadow line would say, "It is dawn, the sun is rising."

The shadow line goes from the North Pole to the South Pole, and for all the people on that line — say, from Norway to South Africa — it is the same time, daybreak. And it will be the same all through the day: all the people along this line where the shadow line was, will continue to have the same time right through the day and night.

At noon all the people along the line where the shadow was will see the sun at its highest point. People further west will see it lower but rising, and people further east see it lower but going down. But all the people on that shadow line have the same time of day.

But one thing is different. We know that the sun is not very high in the north in winter, even at its highest point at noon.

But someone living on the same shadow line, say in Africa, would see the sun much higher at noon. So people on this shadow line all have the same *time* of day, but they see the sun at noon at different *heights*.

We can find these shadow lines drawn on every map. They go from the North to the South Pole, and they are called *meridians* (midday line), for all the people along it have midday at the same time. And the meridians on the maps have numbers. While you could start counting meridians anywhere on the globe, by common agreement the meridian going through Greenwich is the start on every map in the world, and it is numbered 0 (not 1). Meridians are numbered like the degrees on a circle (or on a protractor) going from 0° to 180° east of Greenwich, and from 0° to 180° west of Greenwich. The meridian is called the *longitude* of that place. The longitude of Edinburgh is 3°12' west of Greenwich, or simply 3°12' W. New York is 74° W, and Tokyo is 139°40' E.

But to say exactly where a place is it is not enough to write how many degrees east or west of Greenwich it is. We also need to know how far north or south it is.

The meridian lines all go through the poles like the slices of an orange, they are related to the poles. The other kind of lines must have something to do with the equator. They are lines parallel to the equator and are therefore called *parallels*. The equator itself is starting parallel, 0°, and the other parallels go then to 90° south and to 90° north of the equator. These parallels tell us the *latitude* of a place (which means your distance between equator and pole.)

How will the sun appear to people living on the same parallel, on the same latitude? They will not have the same time of the day; they will all have noon at different times. But when they have noon, the people on the same latitude will see the sun at the same *height*.

We can now give the exact location of any place on the earth by stating the latitude and longitude. This is particularly useful for ships at sea where there are no landmarks to tell where they are. So, for instance, New Orleans lies at 30° N, 90° W; or Sydney is at 33°52' S, 151°12' E.

In summary, the meridian, or longitude east or west of Greenwich, has the same time of day, but the midday sun is at different heights. The parallel, latitude north or south of the equator, has different times, but the midday sun will be at the same height.

23

The Circle

We have looked at the movements of the sun in the sky. Though we see only a part of the path of the sun, we know that this path is a circle. And we have seen that the stars — all the stars in the sky — also move in circles. They move in concentric circles round the Pole Star.

But the sun itself is a circle as we can see it at sunrise and at sunset; the full moon is a circle (even the stars are very small circles). The sun and the moon are themselves circles, and their paths are circles. A rainbow is a part of a great circle: it is as if the sun would paint a picture of itself on rain and clouds.

The ancient Greeks called the great world of sun and moon and stars, "that which makes beautiful," the *cosmos.* And it is still so called today. Cosmos means "adornment," or "beautiful." Out there in the cosmos the lights you see — sun, moon, stars, and their paths — are all what the Greeks called the "perfect form," the circle.

The heavenly circle of the sun also works on earth. If we look at the blossoms of many flowers, we can see that the petals stand in a circle. But in the sunlight the blossoms become fruits and now think how many kinds of fruit are rounded globes — cherries, oranges, grapefruit, coconuts, peas and poppy seeds. If we cut through the stem of most plants we can again see a little circle. Even if we cut through a tree trunk, the rings are not perfect circles, but they are trying to be concentric circles.

If we watch a cat trying to be really comfortable, we can see it rolling itself into a little ball. It tries to make itself into a ball or globe. Snakes, too, roll themselves up into tight, round coils, and birds, when they go to sleep, often tuck their heads in and become round little balls of feathers.

The human head that we carry proudly on our shoulders — upright towards the cosmos, the world of stars and sun — the head too is rounded and, especially the skull protecting the brain, is a globe.

All these living things have rounded shapes, and want to form a ball, a globe. But crystals, for instance, are not alive, and their shape is not at all rounded. They have straight edges and sharp corners. Crystals are not alive, and there is no such thing as a round crystal.

Or the peaks of the Himalayas — these mighty rocks are certainly not alive — have sharp ridges and pointed peaks. Only over thousands of years will wind and weather gradually round them.

Things which are dead and without life tend to be made of straight lines and sharp edges, while what is alive tends to have rounded forms, like circles, and globes. And the rounded shape reflects the cosmos, the world of sun, moon and stars. We can understand that life itself comes from the cosmos. With the light of sun, moon and stars, life also flows from the cosmos.

The round forms, the circles, the globes, are the handwriting of the cosmos which we can recognize just as we recognize someone's handwriting. If we find a round shell that is dead the round form tells us that this was once part of a living creature and its life came from the cosmos.

We can now begin to understand why the Greeks called the circle the perfect form. It is the form of life.

24

The Stars and Sirius

The sun is like a trumpet that calls us to work, to be active; the stars are more like an orchestra that invites us to sit quietly and listen. And people have always felt that in some way they are related to the sun which calls us to our tasks, but there is also something in us that feels a kinship with the stars.

Nowadays, in the age of space exploration, people do not feel this connection with the stars as it was felt in ancient times. If you had told a man from ancient Babylonia or Egypt that we can send a spaceship up to Mars or Venus he would have said, "That's a quite useless exercise because if you patiently watch the movements of these planets, how they grow brighter or dimmer, the stars themselves will tell you all you need to know about them. And it is far more fruitful for the human spirit to reach up to the stars than to send up human bodies, which are not really made for space travel, and have to be put into a strange armour to survive."

Of course we can't go back to the ways of ancient Babylonia, but it is still true that the movements which can be observed are quite mysterious and wonderful, and we shall consider them first.

If we look up to the night sky when it is not cloudy, the stars all look the same, except that some are brighter than the others, But if we patiently watch several nights in a row, we may discover a star which changes its position in relation to the others. That star which has not stayed in its place is actually not a star, but a planet. The real stars are all fixed stars, but the ones that move in relation to the others are planets.

There is another difference between the planets and the fixed stars. The fixed stars shine with their own light but the planets have no light of their own, they only reflect the sunlight which falls on them.

We will first consider the fixed stars, of which there are millions, while there are only nine or ten planets that can be observed from earth.* With the naked eye only about six or seven thousand stars can be seen. With a telescope, the more powerful it is, the more stars can be seen.

From ancient times the fixed stars which are close together have been grouped together and are known by one name, usually taken from mythology. Modern astronomy has kept these old names (as well as adding some new ones), so on a star map we will find a group of stars called Perseus and another called Andromeda. These are taken from the Greek myth. Such a group of fixed stars is called a *constellation.*

One of the constellations which can easily be found in the southern part of the sky in winter is Orion. It can be recognized by three stars in a straight line, the "belt" of Orion. Orion was a great hunter, and below this constellation there are two other constellations, his two dogs, the Little Dog and the Great Dog. The brightest star in the Great Dog is also the brightest fixed star in the sky. It is called Sirius.

This bright star, Sirius, has always been of great interest to astronomers. In the nineteenth century there was a German astronomer, Friedrich Bessel, who made a special study of Sirius. We said the fixed stars do not move. But they do move very, very slightly, but this movement can only be detected with a very big telescope. So Bessel was not surprised that the star Sirius showed a slight movement. But what he observed was not what he had expected. Sirius did not move in a straight line, but in a curve. And Bessel wondered why this was so.

It occurred to him one day that there might be another star so close to Sirius that it had an effect on its movement.

* Pluto, the tenth planet, discovered in 1930, has since 2006 been classified as a dwarf planet, of which currently five are known.

Bessel, who looked at Sirius through a powerful telescope which showed many more stars that you can see with the naked eye, could not find any star close enough to Sirius to influence its movement. But almost twenty years later, in 1862, an American, Alvon Clark, made a bigger and more powerful telescope than any that had existed before. He had a son, a boy of 14, and showed him how to look at the stars through it. The boy turned the big telescope to Sirius and suddenly he shouted, "Father, look Sirius has a little companion."

Through the new telescope they could see the star which Bessel had not been able to find. But Bessel through his thinking had deduced that there must be such a companion for Sirius. The human mind, the human spirit can reach out to the stars.

25

The Daily Movement of Stars and Sun

Of the many constellations in the sky — there are now 88 such groups covering the whole sky — some are more important than others. We mentioned one constellation which was very useful in ancient times, the Plough or Great Bear. From the Plough, which can be easily found in the northern sky, one can find the Pole Star. The Pole Star is the only star in the sky which does not move. All the other stars move in concentric circles around it, or rather, the whole starry sky rotates around the Pole Star. We can observe this if we carefully watch the night sky for a few hours. The constellation Plough swings around the Pole Star, always pointing at it (see page 83). The constellations further away from the Pole star describe big circles, so big that only a part of the circle is above the horizon, and the rest is below where we cannot see it. In one day the Plough makes a complete circle round the Pole Star and, in our northern latitudes, is never below the horizon.

In ancient times people had no watches or clocks; there were hour-glasses, but few people had those. During the day they could estimate what the time was from the position of the sun or, more exactly, from a sundial. And during the night they looked at the position of the constellation of the Plough. The Plough rotates around the Pole Star in approximately 24 hours and, with some practice, it is possible to guess fairly accurately what the time is. Of course, it is not quite exact but most people had no need to know the time exactly to the minute or second.

However, the astronomers of Babylonia, of Greece and Rome, who wanted to be very accurate in their time measurements, discovered something one can also observe today by using a watch. The Plough describes a full circle and comes back to the same position in less than 24 hours. It is only a difference of a few minutes per day but after a week or so the difference is quite noticeable. Our watches and clocks — just like the hour-glasses and sundials of ancient times — take their time from the sun. We divide the time from one sunrise to the next into 24 parts which we call hours. The sun takes 24 hours to come back to the same position. But the fixed stars come back to the same position in less time, in about 23 hours and 56 minutes.

We can see the whole starry night-sky rotating. But this rotation does not stop during the day and when the sun rises and then sets. The sun is really part of the same rotation. But — and this is the difference — when one rotation of the stars is complete, the sun is lagging a little behind and comes back to the same position about 4 minutes later. And, of course, we take our time from the sun and not from the stars.

This time-lag between the stars and the sun means that the sun changes its position in relation to the fixed stars. There are always stars in the sky, even in the daytime, but the stars of the day-sky are invisible because the sunlight is too strong. We can see the stars becoming paler and paler every day at dawn, but they are still there even though we cannot see them. You might think it is a pity that we can never see the stars which are in the day-sky, but we will see them again in six months time.

This is how the people of ancient times discovered that the sun moves, changes its position in relation to the stars: some constellations which are in the night-sky in winter, in summer are in the day-sky and cannot be seen.

If we imagine for a moment that we could see the stars and the sun at the same time — the sun in the foreground and the stars in the background — we would then see the sun standing before a particular constellation of stars. Owing to the time-lag of 4 minutes per day the sun would slowly move through this constellation and after about 30 days the sun would stand in front of another constellation. Wherever the sun is, that constellation as well as the ones next to it are outshone by the sunlight, but the others are in the night sky and can be seen. The constellations through which the sun passes form a circle, and when the sun has passed the last of these constellations it comes back to the first one, just as the hand on our watch comes back to the 12 o'clock and then starts again. Just as there are 12 hours on our watch so there are, for the sun, twelve constellations. The hour hand on our watch makes a full circle in 12 hours, and the sun makes its full circle in 12 months, that is of course, a year.

There are twelve constellations through which the sun goes in one year and that's why we have twelve months in a year. Once, in the past, the months coincided with the passage of the sun through each constellation, but in the course of time people changed the calendar to suit their own purposes without bothering about the position of the sun, and so we cannot say that each month belongs to a particular constellation. But that we have 12 months in a year is due to the 12 constellations through which the sun goes in one year.

26

The Zodiac and Precession of the Equinox

The circle of twelve constellations through which the sun goes in one year is called the *zodiac,* a Greek word meaning the "circle of animals." However, not all the twelve are named after animals. There is a little rhyme which might help to remember them:

The Ram, the Bull, the heavenly Twins,
Beside the Crab the Lion stands,
The Virgin and the Scales;
The Scorpion, Archer and the Goat,
The Man who pours the water out,
And Fish with glittering tails.

They are usually called by their Latin names, and one also often sees the ancient symbols for these twelve constellations: Aries ♈, Taurus ♉, Gemini ♊, Cancer ♋, Leo ♌, Virgo ♍, Libra ♎, Scorpio ♏, Sagittarius ♐, Capricorn ♑, Aquarius ♒, and Pisces ♓.

This "circle of animals" or zodiac is not merely circle but a circular belt. The reason it is important in astronomy is that not only the sun, but also the planets all move along this belt. There are many constellations in the sky, but sun, moon and planets move only within the narrow belt of these 12 constellations of the zodiac.

As well as sun, moon and planets moving through the zodiac, something else also moves through the zodiac, though we cannot see it at all because it is only a mathematical point. It moves so slowly that it takes hundreds of years to notice that it has moved at all. Yet, this movement is very important for the whole of humanity. What is this strange mathematical point?

There are two days of the year when day and night are equal, when there are 12 hours of daylight and 12 hours of darkness: one in the spring and one in the autumn. They are called *equinox* (from the Latin, *equi,* equal and *nox,* night). The equator of the earth has its name because it is the circle where day and night are always equal.

The spring equinox is the one which in ancient times (when most people were working on the land as farmers) was regarded as a very special date in the year. From that day onwards daytime was longer than night, the sunlight grew in strength and warmth, and the seeds the farmers had planted sent out little sprouts. It was the beginning of spring. And in Egypt a sacred white bull, the Apis-Bull, was led through the streets to celebrate this occasion. Why a bull? Because at the time of ancient Egypt on the day of the spring equinox, March 20, the sun stood in the constellation of Taurus, the Bull.

About two thousand years later, in the time of ancient Greece and Rome, people no longer celebrated March 20, the spring equinox. But if they had, it would not have been a bull that was led through the streets, it would have been a ram, because at that time the sun shone from the constellation Aries, the Ram, on March 20. And if we had such a custom today, we would have to carry two fishes through the streets, because now the sun stands in the constellation of Pisces, the Fishes, on March 20.

The position of the sun at the spring equinox, or the *vernal point* has moved through three constellations since the time of ancient Egypt: Taurus, Aries, Pisces. This movement is called the *precession* of the equinox. But why should it matter to us where the sun stands on March 20?

Around AD 1400 a number of great things began to be discovered or invented. About this time the European voyages of discovery began; these led to the discovery of America and culminated in the circumnavigation of the world. New inventions were made, such as book printing, which made it possible for many people to have books and learn about the world. There was also the invention of gun-powder which made knights' armour useless. From that time onwards the flow of discoveries and inventions never stopped. In the 600 years since 1400 more things have been discovered and invented than in all the thousands of years of human history before 1400.

This is not, as one might think, because the Greeks or Romans were not clever enough. For instance, Heron, a clever Greek in Alexandria made a little contraption which used the steam of boiling water to turn a wheel, but this was only a toy to amuse and no one thought that there could be a practical use for this idea. The Greeks and Romans were as clever as we are, but they were not interested in technical inventions. What has changed is human interest.

The interests of the Egyptians and Babylonians were different from those of the Greeks, just as the interests of the Greeks were not the same as ours. Whenever the spring equinox passes from one constellation to another there is such a change of human interest. There are people who say that when the next change comes, when the spring equinox will pass from the constellation Pisces, the Fishes, into the constellation Aquarius, the Water-Carrier, then humanity will become more interested in spiritual matters than in material things, and will have a much stronger feeling that all people are brothers and sisters, and must help each other. So the movement of this mathematical point of the vernal equinox means something for humanity.

27

The Cosmic or Platonic Year

The spring equinox, the point where the sun stands on March 20, moves through the zodiac. While the sun moves round the whole zodiac in 12 months, astronomers have calculated that it takes about 25 920 years for the spring equinox to move around the zodiac. That is a long time, but it is a figure which is interesting for another reason. We breathe at an average about 18 times a minute, and that makes 25 920 times in a day.

The Greek astronomers, who worked out this figure of 25 920 years, called this movement of the spring equinox around the zodiac a *Cosmic Year,* or a *Platonic Year,* after the great Greek philosopher Plato. So, the Cosmic Year has a length of 25 920 ordinary years. So how long is a cosmic month? Dividing 25 920 years by 12, it is 2 160 years. This is the time it takes for the spring equinox to pass through one whole constellation of the zodiac. For 2 160 years the spring equinox is in the same constellation, slowly moving through it. And then it passes on to the next constellation.

In this time human interests change. The Egyptian civilization was in the "cosmic month" of the Bull, the Greeks lived in the cosmic month of the Ram and we are now in the cosmic month of the Fishes.

An ordinary month has 30 days. Dividing the cosmic month of 2 160 years by 30 we get a "cosmic day" of 72 years. A cosmic day is 72 years long which consists of 25 920 ordinary days, which means one ordinary day is one "cosmic breath." Or, in other words, one cosmic breath takes as long as 25 920 human breaths.

Our own breathing and even our life is "tuned in" to the great rhythm of the cosmos. But our breathing is not something separate from our whole organization: our heartbeat is "tuned in" to our breathing. There are four heartbeats to one breath. And as our heart beats, so the blood flows through our body. The most important rhythms of our life are in tune or in harmony with the rhythms of the cosmos. And it is therefore not so strange that a change in the cosmos, for instance, when the spring equinox passes over from the constellation Ram to the constellation Fishes is accompanied by a change in the way human beings feel and where their interests lie.

Another thing which changes with the movement of the spring equinox is the Pole Star, the star which stays in the same place while the whole sky with all the stars in it is turning around it. But the star we call the Pole Star in our time is not the same star as the one which the Egyptians and Babylonians saw as their Pole star.

In the long course of time different stars become the Pole star, and all the stars which, one after another, become Pole stars, lie in a little circle. The time it takes until all the stars in that little circle have had their turn at being Pole stars is 25 920 years.

Why is the Pole star not always the same star? We said the whole starry sky is rotating — that this is how it appears to us — but that it is really the earth which is rotating on its axis; the earth is spinning round. The earth turns round its axis, and this axis points in a certain direction — to the Pole Star. When we see the Pole Star we see the direction in which the axis of the earth points.

If we take an old-fashioned wooden spinning top, and make it spin, it will not remain exactly upright, but will wobble (especially as it slows down). The axis of the earth has a slight wobble, and this means that it points in different directions, but always describing a circle. It is a very slow wobble, but owing to this wobble the axis points to different stars which in turn become the Pole Star.

So the whole earth is involved in this great cosmic rhythm of 25 920 years.

28

The Seven Classical Planets

Where do the names of the days of the week come from? Few people still know that they come from astronomy. Saturday gets its name from the planet Saturn. Sunday of course, is called after the sun, just as Monday is called after the moon. Tuesday is named after the Norse god of war, Tiu (or Týr); the Romans called the same god Mars, and in French the day is called *mardi,* the day of Mars. Mars is a planet. Wednesday is named after the Norse god of the winds and the air, Wodan (or Odin) whom the Romans called Mercury, and in French this day is *mercredi.* It is the day of the planet Mercury. Thursday is named after the god of thunder and lightning, Thor, whom the Romans called Jupiter, in French the day is *jeudi.* It is the day of the planet Jupiter. And Friday is the day of the Norse goddess of beauty, Freia, whose Latin or Roman name was Venus, and the French call this day *vendredi.* It is the day named after the planet Venus.

Saturn, sun, moon, Mars, Mercury, Jupiter, Venus — these are the seven heavenly bodies which can be seen by the naked eye that change their position in relation to the fixed stars of the zodiac. They are the "wandering stars," or *planetei* in Greek from which we have our word *planet.* And these wandering heavenly bodies are always in the constellations of the zodiac.

These seven heavenly bodies, however, move at quite different speeds through the zodiac. The ancient astronomers concluded that the longer a planet takes to make one round of the zodiac, the further away it is from the earth. They thought of the paths of the seven heavenly bodies as concentric circles, which, as we shall find out, is not quite true.

As the seven wandering stars have different speeds the astronomers speak of faster or slower planets, but we must keep in mind that we only notice that a planet is moving by observing it over a longer period. The movements of the planets are like the growing of a plant. It takes some time before you can see that the plant has grown, yet the plant is steadily growing and the planets are steadily moving.

Let us look at how long they take to go round the zodiac. The times given here are approximate, but for just now we are only comparing their speeds.

Moon	1 month
Mercury	12 months
Venus	12 months
Sun	12 months
Mars	2 years
Jupiter	12 years
Saturn	30 years

Mercury, Venus and the sun each take a year to move round the zodiac. While the sun moves slowly and steadily, Mercury sometimes dashes ahead, and sometimes lingers behind — three times a year dashing ahead, and three times a year lingering behind, so you could say Mercury has a kind of movement of about 4 months. Venus does something similar but more slowly. For this reason the Greek astronomers called Mercury and Venus the faster planets, and placed them between moon and sun.

If we look at these times we can see that the moon is 24 times faster than Mars, moon and Mars are like an hour is to a day. Jupiter takes 12 years, in one year it goes only as far as the sun in one month. So Jupiter is to the sun like a year to a month. This is the opposite of the sun-moon relation, where the moon is to the sun like a month is to a year. Saturn needs a long time, 30 years, to complete its round of the zodiac, that is 360 months or almost as many months as the sun takes days.

So Saturn is to the sun like a month to a day. One could also say that Saturn takes as many years as the moon takes days to complete a round. So Saturn is to the moon like a year to a day.

So we can see the paths of the seven heavenly bodies and the pace at which they move are interrelated in a marvellous pattern. Sun, moon and planets are "tuned" to each other, and move in harmony. And as we saw with the movement of the spring equinox, we ourselves with our breathing and our heartbeat are also connected with the rhythms of the cosmos. So the human being and the sun, moon, planets, are all part of the great pattern written in the sky as the constellations of the zodiac.

29

The Moon

The heavenly body which is nearest to our earth, that is the one whose path forms the smallest circle and so completes a full round of the zodiac in the shortest time, is the moon. Now the moon has one thing in common with the planets Saturn, Jupiter, Mars, Venus and Mercury. It is that they all reflect the light of the sun. They have no light of their own and we can only see them because the sun shines on them and we see the reflected light.

With the planets this is not so easy to see but the moon shows it very clearly with its phases: waxing, waning, full and new moon. The bright, lit side of the moon always faces the sun, and during the month as the moon moves around the zodiac, we see the phases changing.

During the course of a month, we first see the fine crescent of the moon in the evening just after sunset. Over the next days it grows — it is waxing — until about a week later it is half moon. The moon continues to grow, to wax, for about another week until it is full. At this phase, at full moon, it is always in the constellation of the zodiac opposite the constellation where the sun is. When the sun is in Aries the full moon is in Libra, and a month later when the sun is in Taurus the full moon is in Scorpio, and so on.

After full moon it begins to get smaller, to wane. The lop-sided shape between half moon and full moon is called *gibbous.* And it continues to wane past half moon, to a waning crescent moon which can be seen in the early morning before sunrise. Then the moon is not visible for two or three days while it

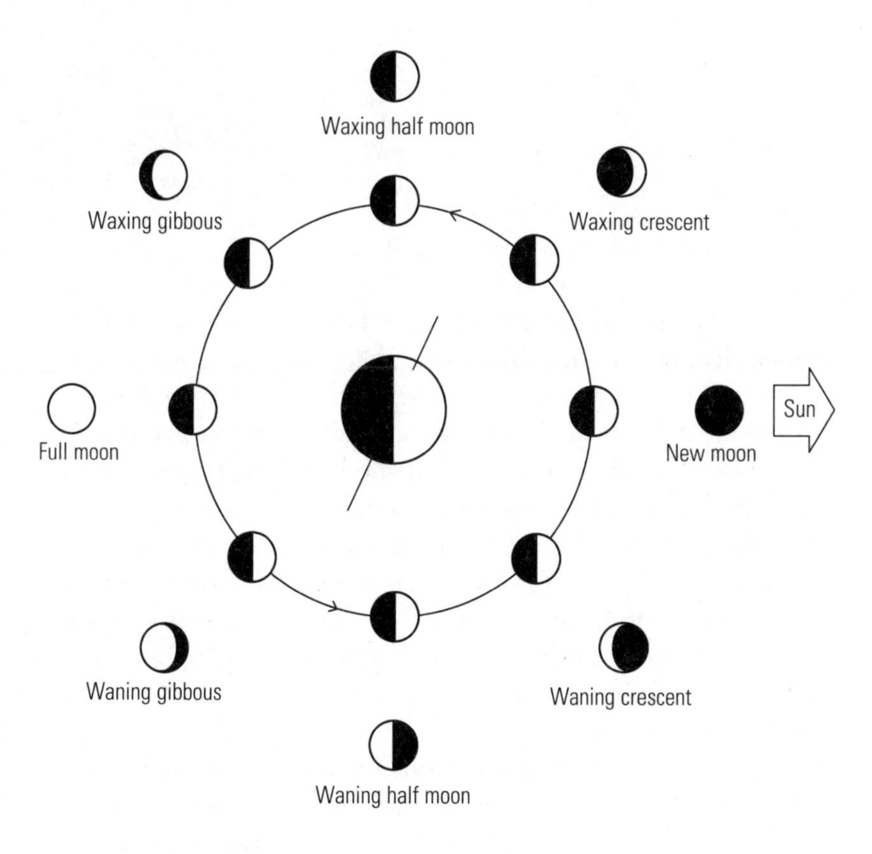

The phases of the moon

passes the sun. While we like to call the thin evening crescent of the moon the new moon, astronomers call new moon the moment the moon passes the sun, just the time when we cannot see anything at all.

Usually the new moon is a little above or below the sun, but once or twice (rarely three times) a year the new moon stands in front of the sun and then the black disc of the moon darkens the bright sun disc either wholly or partially, depending where the moon happens to be. This is an *eclipse* of the sun.

The diagram overleaf shows a fine point of total shadow, the *umbra,* reaching the earth. If we happen to be standing

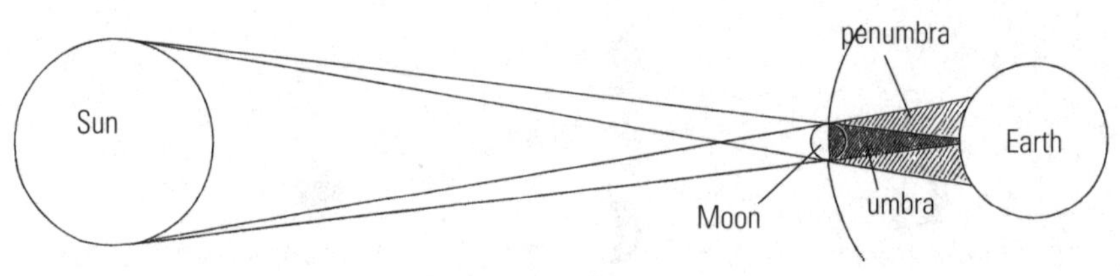

An eclipse of the sun

just in that part of the earth during the eclipse we would be in total shadow: the sun will be completely darkened or eclipsed. Around the dark shadow is another part where there is a partial shadow, the *penumbra.* If we were standing in this penumbra, we would be able to see part of the sun, but part would be blocked: we would see a partial eclipse.

As the moon is passing across the disc of the sun, the shadow moves, and for each total eclipse the shadow moves across the earth from west to east, making a line. So if we want to see an eclipse of the sun, we have to be in the right part of the earth to see it. Astronomers will travel for thousands of miles to observe this.

At full moon, it can also happen that sun, earth and moon stand exactly in line. Then the earth casts its shadow onto the full moon, and there is an eclipse of the moon. There usually are two moon eclipses in a year, and as we can see from the figure, provided the moon is above the horizon (and at full moon that is any time between sunset and sunrise), we will be able to see the eclipse.

Of course, the earth is always casting a shadow but if there is nothing for it to fall on, we cannot see it. But when the full

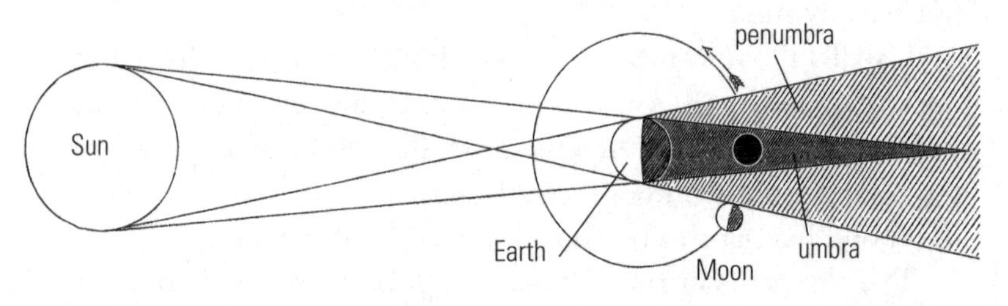

An eclipse of the sun

moon is in line with earth and sun, then the earth shadow does fall on it. At an eclipse of the sun the shadow of the moon falls on the earth. At an eclipse of the moon the earth shadow falls on the moon.

Just as the earth casts a shadow from its dark side, so its light side, the day side, reflects the light of the sun. Again this reflection goes out into space and we do not notice it. But there are occasions when one can notice it. If there is a clear sky when the slender crescent moon appears, we can see the old moon in the new moon's arms. The dark part of the moon is still visible, gently illuminated by the bluish reflected light that comes from the earth, the earth-shine. Springtime is the best season to see it. This is because the new moon in spring stands higher above the horizon than it does at other times.

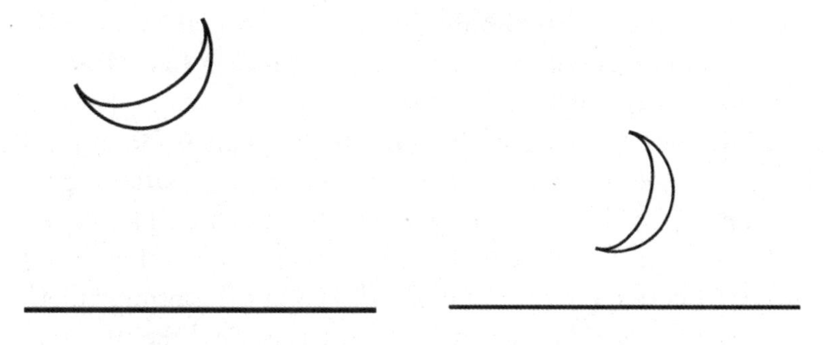

Spring new moon *Autumn new moon*

Both earth and moon receive their light from the sun but, when the moon is new, it is not only that the light crescent of the moon sends a little light to the dark side of earth, the earth also sends its light to the dark side of the moon. It is a time when earth and moon greet each other with light; that's different from the times of eclipses, when they cast shadows on one other. At new moon the dark side of the moon is made a little brighter by the light which our earth sends out into the cosmos.

30

Tides and Other Lunar Influences

The moon moves faster through the zodiac than the sun and its movement is also easier to observe. If we look out of the window at 8 pm and see the moon standing just above the top of a tree, we will not see it there the next evening at the same time. It will arrive there 50 minutes later, at 8.50. And the reason for this is that the moon has changed its position, it has moved in relation to the sun.

There are a great many people who know these 50 minutes by which the moon gets later to the same place without ever looking up to the moon. People who live by the sea know the tides. In most of the seas and oceans there are two high tides every 24 hours. If there is a high tide at 8 in the morning, there is a second high tide at 8.25 in the evening. And the next tide is at 8.50 on the following morning. So between one morning high tide and the next, there are 24 hours 50 minutes. High tide comes 50 minutes later every day: the tides follow the rhythm of the moon.

But the high tides are not always of the same height. The highest high tides and lowest low tide (called *spring tides,* from Old English *springere,* meaning to rise or spring up, not from the season called spring) are always around the time of full moon or new moon. The tides with the least change (around the time of the waxing or waning half moon) are called *neap tides.* So the moon has certain effects on things that happen here on earth.

There are also other and more subtle effects of the moon. Around 1920 a German scientist, Lily Kolisko, planted the seeds of carrots, radishes and other vegetables three days before the full moon, that is at the waxing phase. And she also planted the same kind of seeds three days before the new moon, at the waning phase. The plants of the waxing phase did much better, grew faster and were stronger than the plants of the waning phase. Since then many gardeners and farmers all over the world have followed the advice of this scientist and have profited from it.

There are other and even more subtle effects of the moon. Nurses who work in hospitals for people who are mentally ill know that at the time of the full moon many of their patients become much worse and are more difficult to handle. This was so well known in ancient times that madmen were simply called "lunatics" which comes from the Latin word *luna,* the moon. A "lunatic" meant someone who reacts too strongly to the influence of the full moon. But there is also a good side to the mental influence of the full moon. Writers, poets, composers of music, people who need a great deal of imagination for their work have more ideas at the time of the full moon than at other times.

The peasants in the Middle Ages always reckoned that rain was more likely at the time of the full moon or the new moon. Then came the scientists and said it is all superstition. But some years ago an Australian meteorologist, Bowen, collected the frequency of rainfall from places all over the world and over long periods and found that the stupid peasants had not been so stupid and that rain is indeed more frequent at the times of the full or the new moon.

There are many examples of the influence of the moon in the animal world. One of the strangest instances is the palolo worm, a sea-worm that lives in the Pacific Ocean. This worm is regarded as a delicacy by the natives of Samoa, but they can enjoy eating their fried palolo worms only once a year. Just at dawn one week after the full moon following the equinox, the worms come out of the deep sea where they live, to spawn their eggs. They come up in their millions and are caught by the Samoans. No one knows how the worms know the phase of the moon, for the full moon does not occur on the same date every year. But whatever the actual date is, on that day the worms come up from the depth. They cannot see the moon in the dark depth of the sea where they live, so it is a mystery how they time their spawning so exactly.*

As there are so many strange influences connected with the moon it is not surprising that there are many fables and legends about it. There is a very beautiful one which comes from Africa.

When we in Europe look up at the full moon we see the dark patches as a kind of face and we call it the "man in the moon." But to the Africans these patches looked like an animal, like a hare, and they tell a story how the hare got up there. They say that in very ancient times human beings could see that when a person died the soul went up to God in the light of the sun and the moon and the stars, they could see it as we see clouds or the rainbow in the sky. And because they could see what happened after death they were not afraid of dying. Death was for them no more than a journey to another land. But as time passed men lost the ability to see where the souls of the dead went and so they began to be afraid of death, they feared death.

And the moon-spirit took pity on these poor people on earth. He said to the hare, "You shall be my messenger. Go to the men on earth and tell them to look at the moon. Just as the moon gets less and less after it has been full and disappears at the new moon, but then appears again as a thin crescent which grows and grows, just like that the souls of the people who have died and disappeared will come and live again. The new moon is not the end of the moon-existence and death is not the end of human existence."

* L. Kolisko, *The Moon and the Growth of Plants,* Anthroposophical Press, London 1938. Rainfall and phases: E. G. Bowen, reported in *New Scientist,* 7 March 1963. Palolo worm: Ralph Buchsbaum, *Animals Without Backbones,* USA 1938.

The hare obeyed and went to the men on earth to bring them the message of the moon. But the men did not listen: they hunted the hare, shooting arrows at it, and chased it with their dogs, so the hare ran away without delivering the message. And the moon spirit was angry with the hare and threw it onto the moon where the Africans can still see it today. But how did people then know what the moon spirit had said to the hare? The African tribes say that the moon whispered its message to good and wise men in their dreams. As there are not so many good and wise men about, there have always been only a few to whom the moon could speak in their dreams, and they told the others.

31

Easter

Spring is the time of the vernal equinox, the time when neither day nor night are stronger, each lasting exactly twelve hours. At the time of the spring equinox the sun shines down from the direction of the constellation of the zodiac which is as important for our time as the constellation of the Bull was for ancient Egypt, or the Ram was for Greece and Rome. In our time the constellation where the sun stands at the time of the spring equinox is the Fishes. Fish move freely in the oceans, they range far and wide, (think of the salmon which swim thousands of miles from the Atlantic Ocean and up our rivers). Our time is a time when people should range far and wide and explore the world, also with their mind and spirit. That is the message which the sunlight brings to us at the spring equinox, March 20, when the sun stands before the constellation of the Fishes.

March 20 is also called the beginning of spring in our calendars (of course, it is different in the southern hemisphere). It is the time when nature all around us is growing and budding, when the cold days of winter are past and there is new life everywhere. In the African story we heard how people on earth were told to look up and to learn from the waning moon that the new moon is not an end, but will be followed by the waxing moon, the sign of new life. Before Easter not only is this waxing half moon highest in the sky, but here on earth too all things are waxing, that is, growing. Life here on earth also tells us: in winter trees may look dead and barren, the fields may look dead and bare, but it only seems so, for now there is new life everywhere. So at Easter nature around us, and the moon in the sky, the sun and the stars, they all have a special message for us at this special time of the year.

In the African story, in a time long past people had no fear of death for they could see the souls of those who died rise up in the light of sun, moon, stars. But later they lost this ability, and could no longer see what happened after death; they lived in fear of death. In the times of ancient India, as we heard in the story of the Pandu brothers, people even looked forward to death, they were glad to leave earth. Gilgamesh, who lived four thousand years later, was already afraid of death.

It was to show us that death is not an end but the beginning of a new life, that Christ went through death and rose from the grave. That is what happened at the first Easter.

So why is Easter not on the same date every year? The early Christians said, "The death and resurrection of Christ is not only something that concerns people on earth, it is a great event for the cosmos, and when we celebrate Easter we want to take account of the cosmos, of sun and moon and stars."

So they made the following rule: Easter Sunday is the first Sunday after the full moon following the spring equinox. Sunday means the day of the sun, so that takes account of the sun. After the full moon, that takes account of the moon. And after the spring equinox, takes account of the stars where the sun is at the equinox.

That is why Easter is not always on the same date; Christmas is always on December 25, but the date of Easter changes from year to year. It can be as early as March 22 or as late as April 25. Easter is the festival of life, the triumph of life over death, it is a celebration of all the new life that comes with the spring, and all life is connected with the rhythms of the cosmos.

32

The Planets

We now come to the other heavenly bodies that move through the zodiac, the planets Mercury, Venus, Mars, Jupiter and Saturn. The most brilliant of these planets is Venus. Venus is a planet which is never very far from the sun; it either appears before sunrise in the morning, then we call it the morning star, or it follows the sun and shines after sunset in the evening sky and we call it the evening star. Both the morning star and the evening star are the same planet, Venus. Of course Venus is a planet, not a star, it has no light of its own and, like the moon, only reflects the sunlight. (The planets Mercury and Venus, like the moon, show phases, but they take much longer than the moon, and we need binoculars or a telescope to observe them.)

There are also times when Venus is not visible because it is then so near to the sun that its light cannot be seen in the daytime sky. There is a certain rhythm in this. You see Venus for a time as morning star; it gets nearer and nearer to the sun and becomes invisible in the sun's bright light. And then it reappears as evening star. Then the evening star gets nearer and nearer to the sun and disappears, and then reappears as the morning star. When one heavenly body passes another, it is called a *conjunction.* So Venus has a conjunction with the sun after being the evening star and another one after being the morning star.

If we make a drawing of the zodiac as a circle and connect the points where Venus has a conjunction after being the morning star, we find it makes a regular five-pointed star. And

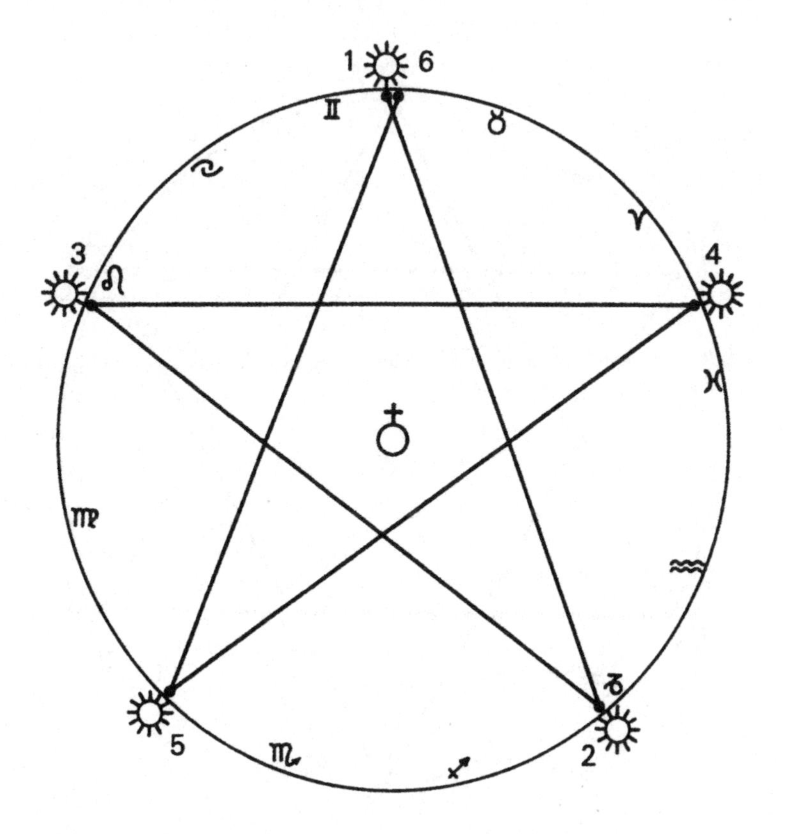

Position in the zodiac of conjunctions of Venus with the sun over eight years (from Schultz, Movement and Rhythms of the Stars*)*

it is the same with the conjunctions which come after Venus has been the evening star. Venus draws five-pointed stars, but you have to be an astronomer to become aware of it.

Just as Venus is the most visible planet because of its brightness, so Mercury is the least visible. It is not just that Mercury is less bright; it is closer to the sun than Venus, and so the light of Mercury can only rarely be seen, particularly the further away from the equator we are. But, like Venus, it has times when it comes up before the sun as a morning star and times when it sets after the sun and is an evening star. And in between it has conjunctions with the sun when it is completely invisible. And if we draw the Mercury-sun conjunctions on a zodiac circle, as

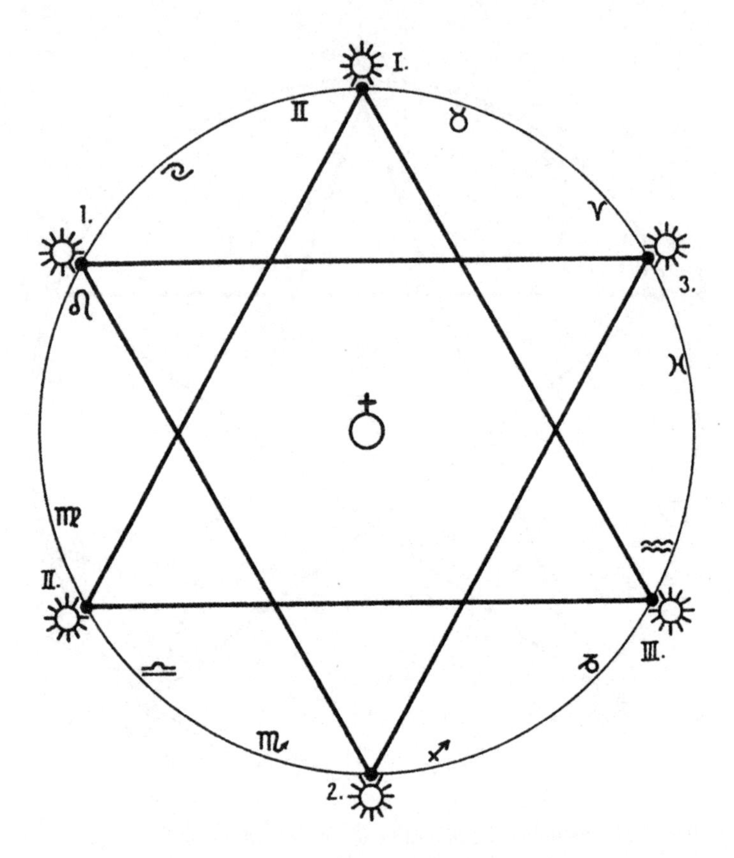

Position in the zodiac of conjunctions of Mercury with the sun over one year. Each triangle connects a similar type of conjunction.
(from Schultz, Movement and Rhythms of the Stars*)*

we did with Venus, then a triangle appears. Mercury and Venus move faster in the zodiac, though as they are "tied" to the sun (always being close), on average they take the same time as the sun — one year.

The planets Mars, Jupiter and Saturn take longer than the sun to travel through the zodiac. Their movement is not tied to the sun's and so they can be high in the midnight sky (which never happens with Venus or Mercury). They are in the night sky when the sun is on the opposite side of the zodiac to where they are. They can also be in the day sky, invisible, which happens when the sun overtakes them, when they have a conjunction with the sun. We can mark these conjunction-points as we did with Venus or Mercury, and we find that they also make interesting patterns. Mars takes about two years between conjunction, and has seven irregularly spaced conjunction points in 15 years. Jupiter makes eleven neat, evenly-spaced conjunctions in 12 years, and Saturn takes 30 years to make 29 conjunctions.

In ancient times astronomers noticed another thing with these slow planets. Carefully observing the position of one of these planets night after night, they discovered that these planets move not only in one direction as the sun and the moon do, they sometimes slow down, stop and go backwards, and then after a time stop again and move forwards. You have to observe them for weeks and months and mark their positions against the background of stars to see how these slow planets make U-turns and loops. We shall hear later how the astronomers tried to explain this movement of Mars, Jupiter and Saturn.

Of these three, Jupiter is the brightest, but not as bright as Venus. Mars is also quite bright and shines with a reddish light. Saturn has the weakest light of them, but can still shine more brightly than most of the surrounding stars.

Sun, moon, Mercury, Venus, Mars, Jupiter, Saturn were known to the astronomers of ancient times in Babylonia and Egypt as the seven wanderers, or seven planets. Of course these astronomers had no telescopes. Since the invention of telescopes two more great planets have been discovered: Uranus and Neptune. These "new" planets are very far away, as we can see from how long they take to go round the zodiac. Uranus, takes almost 85 years and Neptune takes 165 years. They are planets which, even if you observe them the whole year, hardly seem to move at all. Pluto, the tenth planet, was discovered in 1930, but in 2006 astronomers decided it was too small to be a real planet, and it was reclassified as a dwarf planet, of which currently five are known.

33

Pythagoras

Before going on to what is known about the stars, let us look at the great men who have built up astronomy as it is today. The story of astronomy began in Egypt and Babylonia and the names of the priests and wise men who studied the skies patiently over long periods have not been recorded. The first known astronomers lived in ancient Greece. One of the earliest great men who helped to build up that wonderful treasure of knowledge which we have today was Pythagoras. He was born on a little island called Samos in the Aegean Sea around 500 BC (at the same time as Buddha lived in India). His father was a merchant and when the son was born, the father, as was usual in those days, sent messengers to the famous temple of Delphi to ask the oracle what future his newborn son would have. And the priests of the temple answered, "As long as men live on earth the name of this child will be remembered."

As this boy grew up he showed a great love for doing geometrical drawings, and doing sums of any kind was not "work," but a joy and pleasure for him. By the time he was twenty-one he had learned as much of arithmetic and geometry as anybody in the whole of Greece could teach him. But Pythagoras wanted to know more. It was not just ordinary curiosity, there was something more. He had often watched when one of the beautiful Greek temples was built and he had seen how carefully and accurately each pillar, each step, each part of the temple had to be measured and made to just the right size. He had seen the architects working with compasses and rulers and making long calculations, so that all parts of the temple would fit together perfectly. And Pythagoras said to himself, "Without using numbers no temple could be built, nor even the houses people live in. But is not the whole world the temple of God? Surely, the whole world is created and built in accordance with numbers and the more I can understand about numbers and figures, the more I can understand of the world."

Numbers, arithmetic, algebra, geometry, all this is called "mathematics," and mathematics is a way to understand the divine wisdom in all creation, that is why Pythagoras was so anxious to learn more about it.

There was no one in Greece who could have taught him anything he did not know already, but he heard that there were priests in Egypt who knew much more than the Greeks. So Pythagoras left the Greek island of Samos where he was born and sailed to the land of Egypt.

When he came to the wise priests of Egypt and told them he had come to learn from them, they answered: "It is not our custom to share our knowledge with any person who comes along. We regard knowledge as something holy, something sacred, and only people who are worthy should have knowledge. If you want to learn from us you must first prove to us that you are the right person to share our secrets."

In our time we teach science, maths, all subjects, to anybody and everybody. But in ancient Egypt all knowledge was holy and was given only to special people. To prove that he was worthy Pythagoras had to pass certain tests. They were not at all like the tests we know, tests with pencil and paper; they were quite different. For instance, he was given a task that was very dangerous; to show that he had courage. At other times he had to go a long time without food or drink, to show that he was master of his body, not the body master of him. At other times he was not allowed to speak for many months, no matter what happened or what he wanted, to show that he was master of his tongue. And there were other tests. Only when Pythagoras had passed these tests was he accepted as a pupil by the priests of Egypt. He spent many years with them, learning the great wisdom and knowledge they had.

While Pythagoras was in Egypt, the land was invaded and conquered by the Persians. The Persians took many of the Egyptian priests as prisoners and sent them to Persia. For them Pythagoras was also an Egyptian priest, and so he too came as a prisoner to Persia. But the King of Persia had a Greek physician, a doctor, and when this doctor saw a fellow Greek amongst the prisoners, he pleaded with the King, and Pythagoras was set free but had to stay in Persia. Now at this time Babylonia was also under Persian rule and Pythagoras went to Babylonia to learn about the stars and sun and moon — for at that time the priests of Babylonia knew more about the stars than anybody else in the world. Just as Pythagoras had passed the tests of the priests in Egypt, so now again he had to pass hard and dangerous tests until the Babylonian priests took him as a pupil to learn their secrets.

Just as in the Old Testament, the Children of Israel were first in Egypt and then, centuries later, were captives in Babylonia, so it was with Pythagoras. He also met some of the holy men and prophets of the Israelites there and learned from them. After many years Pythagoras was allowed to return to Greece.

But where should he go? In all these long years of his absence his parents had died, he had no relatives and the island where he was born had been conquered by the Persians. But he now possessed more knowledge, secret knowledge, than anybody of his time. Where would he find pupils worthy of sharing his knowledge?

At that time there were Greek cities not only in Greece but also in the south of Italy. For instance, Naples is a Greek name *Neopolis,* meaning "new city." And Pythagoras went to one of these Greek cities in Italy, called Croton.

Pythagoras did not look like an ordinary man. It was not only that he was tall and dressed in the pure white linen of the Egyptian priests, it was not only that his long hair and beard had turned white in these years in foreign lands and by the hardships he had gone through, but in his dark eyes one could see the power of knowledge.

He started a school or college at the Greek town of Croton in southern Italy. What was taught at the college was kept strictly secret and it was only after Pythagoras had died that the secret knowledge — or some parts of it — became known in the world. It was in this way that the Greeks heard for the first time something which other nations had known for a long time. The Greeks at that time had only a very sad idea of what happened to a human soul after death; they thought the souls of the dead lived in a dark world of shadows, the realm of Hades. But Pythagoras told his pupils what he had learned from the wise priests of Egypt: that the souls come back to earth again and are born again. In the Orient, in India, in Egypt, this was nothing new, but in Europe it was Pythagoras who was the first to tell his pupils about *reincarnation,* as it is called.

That was one kind of knowledge Pythagoras brought from the East. Another kind of knowledge was about numbers, but it was quite different from the way we usually think about numbers. When Pythagoras had passed his tests and trials in Egypt, the priests of Egypt told him that the first numbers are not just figures, they have meaning. For instance, the first number, *one,* we might think is a small number, less than 2 or 3, but *one* is really the largest number, for the whole world is one, and all the many things: stars, planets, earth, men, animals, they are parts of the one great world, the universe, and the word "universe" comes from *unus,* which means one. The figure *one* means the whole universe.

Two means all the things which exist in twos: day and night, men and women, love and hate, good and evil. If there were no "two-ness" in the world, there would be no difference, no contrast, and all things would look the same.

Three is related to all things that come in threes: father, mother, child; light, darkness and colour (for the colours are mixtures of darkness and light); waking, dreaming and sleeping.

Another threefoldness is this: in our head we think, in our hearts we feel, with our hands and legs we do things. Human life exists in these three things: thinking, feeling, doing.

Four stands for all "four-ness" in the world. The cardinal points, east, west, south, north; the four seasons; the four kingdoms of nature — man, animal, plant, mineral; the four elements of fire, air, water, earth.

And then the priests said, "Look at the great pyramid which we have built, if you approach it from afar, you see first the peak, the point in which all sides come together. This point stands for one. If you come nearer and see the pyramid from a corner, you see two sides, and one is always darker than the other (because of the sunlight), this view shows the two-ness. But if you see the pyramid from the side, you see the triangle that stands for three. And if you could see the pyramid from above you would see the square base which stands for four-ness in the world."

And then the priests told him that the triangle stands for all threefoldness in the world and the square for all fourfoldness. And there are different kinds of triangles: there are as many kinds of triangles as there are "elements" in the world — four. There is one kind of triangle where all sides are of different lengths, it is called the scalene triangle and it is the triangle of air. There are triangles which have two sides equal, they are called isosceles triangles and these are the triangles of fire. There are triangles where all three sides have the same length, equilateral triangles, they belong to earth. And there is still one more kind of triangle, its sides are of different length but it has one right angle: it is the right-angled triangle, it is the triangle of water. The triangles exist in as many kinds as there are elements.*

Only two of these triangles can be mixed together to make another, new triangle. The triangle of fire (isosceles) and the triangle of water (right-angled). You could say that the fire-

* Taken from Lancelot Hogben, *Mathematics for the Million,* 1936.

triangle is the father and the water-triangle is the mother, their child is the right-angled isosceles triangle.

It has two equal sides and one right angle. This is a very special triangle. If you have two triangles of this kind (and of the same size) and put them together on their base-line they form a square. And if you take four such triangles and put them together at their apexes they again form a square, a square double the size (area) of the first square.

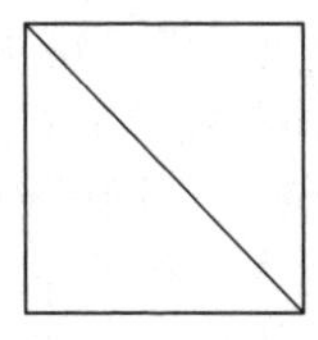

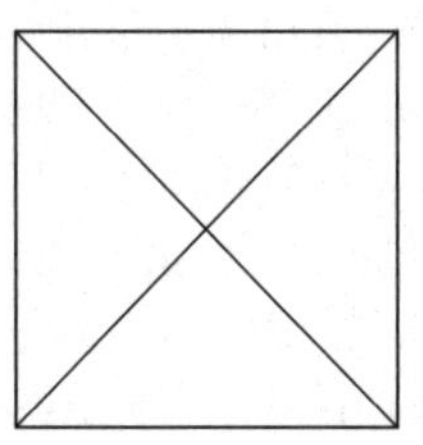

34

Ptolemy

Pythagoras had learned geometry from the priests of Egypt. He had gone further than the Egyptians and had discovered the "theorem" about right-angled triangles which made him famous for all times. It was a simple construction — but because of it, the prophecy of the Oracle in Delphi came true and his name is still remembered today.

When Pythagoras was in Babylonia (then part of Persia), he came to the priests of Babylonia. These great astronomers had gathered more knowledge about sun, moon, planets and stars than any other people of the world.

Many years later Pythagoras spoke to his pupils in Croton about his stay in Babylonia and what he had learned there. He told them how, by some means which remained secret, he was put into a kind of sleep that was not like ordinary sleep and not like an ordinary dream, because his mind was wide awake, but his body was lying almost as if it were dead. His mind was not in the body at all; he felt that he was soaring higher and higher. He was a spirit, free of the body, and rising from the earth into the cosmos. And, rising as a free spirit into the height, he *heard* the cosmos. The sun was a mighty voice, yet more beautiful than any human voice. The moon was a gentle, sweet voice, and the planets had each their own sound. And from the fixed stars far away there came a sound like a great choir; no music on earth can compare with the mighty, ever-changing music of the cosmos. It was not only the planet that we see that sounded, but the path of every planet is part of a great sphere, and the whole sphere of every planet sounded and sang and sent forth music. That was the great "music of the spheres" that can only be heard by souls free of their bodies. But then as he felt himself drawn back to earth, the music of the spheres became fainter; silence and darkness was around him, and then his strange sleep came to an end. He could feel his body again, and awoke.

Pythagoras told his pupils about the "music of the spheres," the music of the cosmos. He also told them that human souls have a very faint memory of this great music they heard before they were born. That is why on earth they make music and enjoy it; they are trying to remember the music of the spheres. And, of course, some people remember this music of the cosmos better than others.

It was a great secret which Pythagoras told his pupils, but this secret was betrayed, and ever since people used to speak of the "harmonies" of the spheres. Even centuries later, Shakespeare, for instance, spoke of it in his play, *The Merchant of Venice,* and there is a poem by Goethe that also speaks of the "music of the spheres."

Pythagoras also told his pupils something they could hardly believe. He said, "You think the stars, the whole sky with stars and sun and moon are turning around you. But you are mistaken. It is the earth which turns round, the earth is a globe which turns round its axis. The axis of the earth points to the Pole Star. That is why it does not appear to move. But the stars, the fixed stars do not move at all, it is the earth which turns round. And the earth does not just turn round on one spot in the cosmos, it moves on a path around the sun."

That was a difficult thing to understand for Pythagoras' pupils. They found it hard to imagine that the solid earth on which they walked was not only spinning round, but moving on a path, on a circle, in the cosmos. Later, when the secrets of Pythagoras were made known, that was one thing few people in Greece or, later in Rome believed. There were a few who thought Pythagoras was right, but most people would not believe it. And so, the name of Pythagoras was praised for his geometry, for this theorem, and the "music of the spheres" was also accepted, but what he had said about earth and its movement was not taken seriously for two thousand years.

About six hundred years after Pythagoras, in Alexandria in Egypt a Greek called Ptolemy wrote a book about astronomy. Ptolemy had heard about Pythagoras and of the spheres and their music. In his book he described that the earth was in the centre of seven spheres which carried the sun, moon and planets, and an eighth sphere which held all the fixed stars. The moon, being the fastest, had the smallest sphere. Then came the spheres — ever larger — of Mercury, Venus, the sun, Mars, Jupiter, Saturn, and finally the greatest sphere was the sphere of the fixed stars.

Pythagoras had also said that the earth moved, but Ptolemy had either not heard that or did not believe it. In any case, he did not mention it in his book. People thought Ptolemy was a very clever man. He was greatly respected in his own time, and for many centuries later.

In time, people no longer understood that these spheres were not something hard and solid; they thought that spheres were like transparent solid globes around the earth and that the planets, sun, moon, rolled around these globes. They imagined the spheres were really something solid which one could touch if one could fly high enough. But Pythagoras had spoken of something he had only known and heard when his soul was out of his body, as long as we are in the body these spheres cannot be seen or touched.

For fifteen hundred years the system of Ptolemy was looked upon as something one could not doubt at all. Only when the spring equinox entered a new constellation, the Fishes, did human minds change and people began to question whether Ptolemy had been right.

35

Copernicus

We have heard about the threefoldness in human life: we *think,* we *feel* and we *do* things. But we are different from one other. Some people are very good at thinking, but when it comes to making anything with their hands they are clumsy. Some people have very warm, kind feelings for others, they will do anything to help others, but they are not very clever when it comes to thinking. And some people are born to do great deeds, like Alexander the Great or Julius Caesar, but they are neither great thinkers nor specially kind hearted.

Alexander or Caesar, Richard the Lionheart or the Vikings who sailed to Vinland were heroes of deeds, not of thinking or feeling. St Francis of Assisi was a hero of feeling, as was Dr Barnardo who had this great love for the poor children without proper homes. There are also heroes of thinking. And Pythagoras was such a hero of thought. The theorem which is named after him is a thought; so was the idea of reincarnation and what he said about the harmony of the spheres or the movement of the earth. These were also great thoughts.

However, the astronomer of Alexandria, Ptolemy, was not a great hero of thought. He only repeated something Pythagoras had said. But for nearly two thousand years Ptolemy ideas were believed by everybody. Soon after the year 1400 the spring equinox came into the constellation of Pisces, the Fishes, and the human mind began to change. After a hundred years, by about 1500, people began to ask questions they had never asked before.

In Poland there lived a canon of the Church, a kind of

priest, whose hobby was astronomy. The name of this priest was Nicolaus Copernicus. Being an educated man, Copernicus read Latin and Greek, and in some books of ancient times which had survived he read about Pythagoras. Pythagoras had not only spoken of the harmonies of the spheres, but had also spoken of the movement of the earth. Now one thing was very complicated in Ptolemy's world picture, and that was the "dance" of the planets. If you observe the planet Jupiter or Mars, for instance, you can see over weeks and months that the planet moves for a time in one direction, but then it moves in the opposite direction making a loop.

The Greek astronomers and Ptolemy explained this by saying the planets moved on a little circle or sphere, the *epicycle,* which moved around the great circle or sphere. Each planet (though not sun and moon) had its great circle as well as its smaller epicycle.

Now Copernicus thought, if I ride on a horse and I overtake another rider on the road, it may seem to me that this

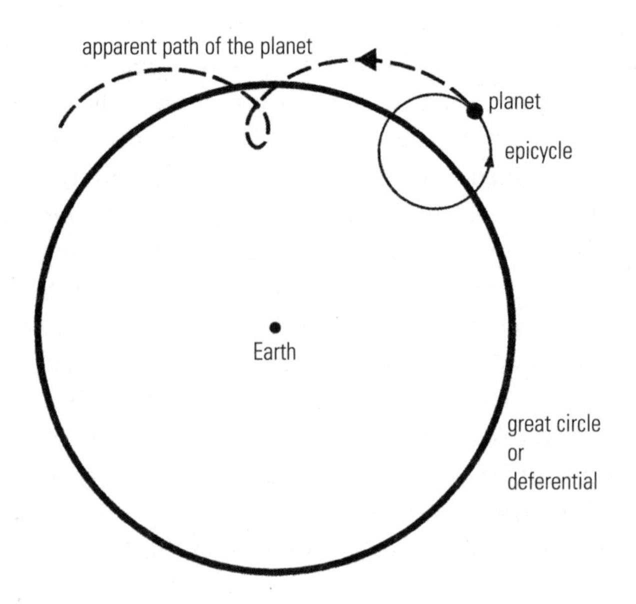

The Ptolemaic system of a planet's great circle (deferential) and epicycle

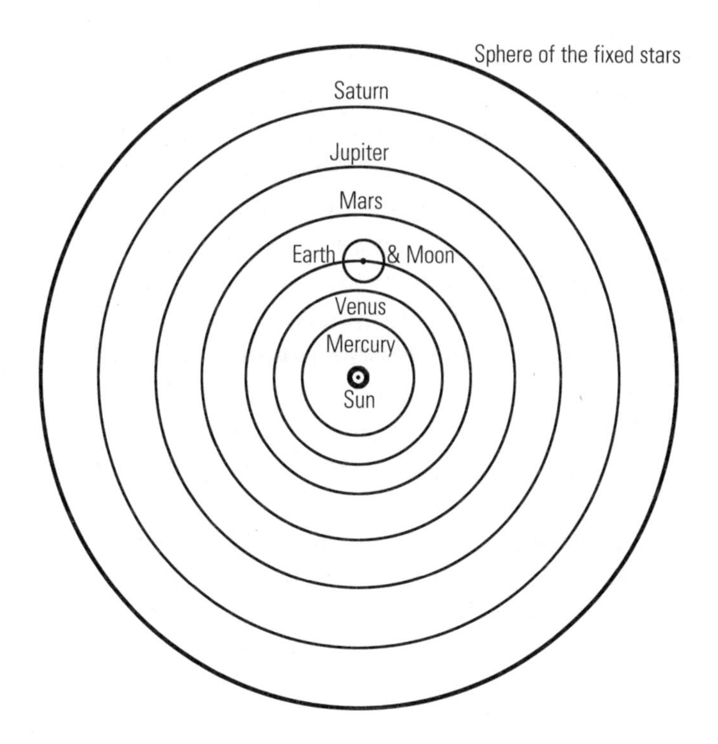

The Copernican system with the sun in the centre

other rider is going backwards, but he is not, it is only my own movement, my own speed, which leaves him further and further behind. Of course, because we see the ground we know the other rider is not really going backwards; but what if we could not see the ground? Perhaps the planets did not really travel backwards, but earth moved on and this made the planets *look* as if they were going backwards. Copernicus found that if he let the earth move around the sun, and the planets too, move around the sun, then because of these two movements the planets *appear* to make loops and go backwards.

So Copernicus said that the sun does not move in a circle round the earth, but that it is the earth which goes in a circle round the sun; the sun must be fixed. And as the earth goes round the sun in one year, we see the sun against a different background of stars, the constellations of the zodiac.

What about the moon? The moon does not make loops, but we see the moon in different constellations of the zodiac during the course of a month. The moon is the only thing moving in a circle round the earth. So the world-picture Copernicus worked out had all planets going round the sun, and the earth also going round the sun, making the earth a planet. But the moon goes round the earth, and so is not a planet, but a *satellite.*

All this was thought out by this man Copernicus, and it gave a much simpler explanation of all the movements we can see in the sky. But there was nothing to prove that it was true, it was all just an idea, a theory, and you can work out the movements of the sun, moon and planets with either system. So for a long time he did not even write a book about his idea. Only when Copernicus was an old man did his friends, whom he had told about his idea, press him to write it all down. In 1543 he finally did, calling his book, *On the Revolutions of the Celestial Spheres.* He saw the first copy of the book on the day he died.

36

Tycho Brahe

Three years after Copernicus died, Tycho Brahe was born. He came from a noble family, his father and uncle were at the court of the King of Denmark. For most of his childhood he lived with his uncle. When Tycho was thirteen years old he was studying at the University of Copenhagen. In those days boys often went to university at the age of 13 or 14. There he heard that there would be an eclipse of the sun the following day. Tycho eagerly watched the next day, and the sun really did turn dark. Young Tycho cried out, "A man who can predict such a thing must be like God." From that day onwards he was determined to become an astronomer.

His uncle wanted Tycho to study law, but he was much more interested in astronomy. At night when he was supposed to sleep he stole out into the garden to watch the stars. He soon knew every star one can see with the naked eye. He studied every available map of the stars. At the age of sixteen he was sent with a tutor to finish his studies of law at universities in Germany. His tutor, however, soon found he had undertaken a hopeless task. However hard he tried, he could not succeed in interesting Tycho in anything but astronomy. Any money Tycho managed to get was immediately secretly spent on astronomical books and instruments. He wrote, "None of the charts match the others. There are as many measurements as there are astronomers and all of them disagree."

He resolved to systematically observe the skies and make the most accurate map of the stars. He made himself special instruments to measure the angles between stars. To the disappointment of his uncle he refused to go riding and hunting as all other young noblemen did. He said, "I prefer to look at God's beautiful work, I am not going to waste any time riding or hunting." When he was nineteen Tycho's uncle died, and he was no longer forced to study law and expect to do things he did not want.

At that time all people — and all astronomers — believed that the stars ruled human life on earth. If you knew the position of the planets on the day a person was born, you could foretell what was going to happen to that person. Now young Tycho wrote a poem in which he predicted that, according to his calculations of the stars, the Sultan of Turkey would die in October 1566 during an eclipse of the moon. And, strangely enough, the Sultan did die around that time. This made Tycho's name known.

However, what made Tycho really famous in the whole of Europe was something that occurred a few years later. One evening in early winter, as he was walking home, he looked up at the sky and saw a brilliant star high in the constellation of Cassiopeia. He immediately knew there was no star there, and he also knew that no planet could ever stray that far from the zodiac. To make sure he called his servants to ask whether they could see this star. They certainly could, and so Tycho knew it was not a creation of his mind, or something he had eaten or drunk.

Tycho immediately when he got home pulled out his great instruments to measure the exact position of the new star. He soon calculated it was not some bright light in the clouds, but a real star, as distant as sun, planets and stars. Having observed the new star for some weeks, Tycho wrote up his observations. His friends tried to dissuade him from writing, pointing out it was beneath the dignity of a nobleman to write a book. But Tycho was undeterred, and published his book, *De Nova Stella.* Like all learned books of that time it was written in Latin. Astronomers still call a new star a "nova" from the Latin.

Following this, the King of Denmark made Tycho Brahe his court astronomer. He gave him a whole island on which

Tycho built the greatest observatory that had ever been built for watching the stars. He had many servants and assistants — instrument-makers, mathematicians, builders. He not only had farms and bakers to supply all the people with food, he even built a paper mill on the island to ensure there was enough paper to write down all his observations. Here on this island Tycho made finer, more exact instruments for mapping the course of the planets than anybody had ever had before, and produced more accurate charts of the stars than ever before.

Tycho, like all astronomers of that time, had been taught the system of Ptolemy with the earth in the centre and the sun, moon and planets going round the earth. But he had also read

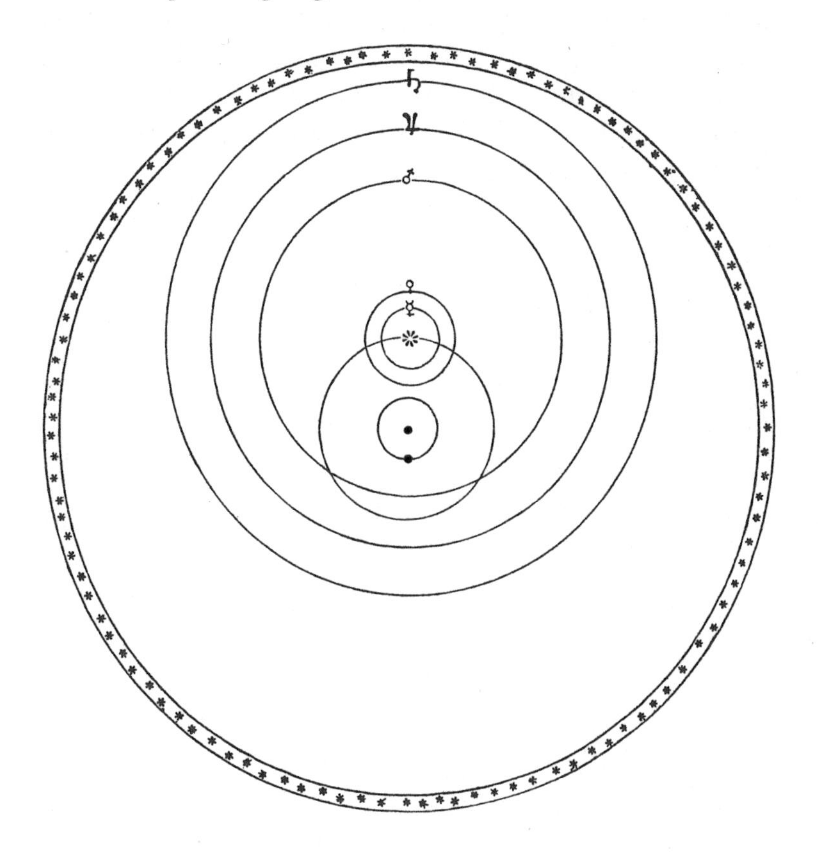

Tycho's system with the earth in the centres, but the planets revolving around the sun

the book by Copernicus with the new idea that the sun stood still and the earth and other planets moved round it.

The system of Ptolemy had one great advantage. It was what we really see: sun, moon and planets going round the earth. The Copernican system, with earth and the planets going round the sun and the moon going round the earth, was simpler, but it was not what we see.

Tycho of course knew both systems, but he did not like to give up the world-picture of Ptolemy altogether. After all, we *do* see the stars and sun and moon going round the earth. And so he thought out a different system. He said the planets — Venus, Mercury, Mars, Jupiter, Saturn — go round the sun (so Copernicus is quite right), but the sun goes round the earth, and so the planets, following the sun, go round the earth too (and so Ptolemy is also right). And the movements of the planets can just as well be worked out according to the world-picture or system of Tycho Brahe.

37

Johannes Kepler

When the King of Denmark died, Tycho quarrelled with the new King. Luckily, Tycho was invited by Emperor Rudolf of Austria to come to his court in Prague as court astronomer.

Meanwhile, a young German, Johannes Kepler, had worked for some years in Austria as professor of astronomy. Kepler thought that the earth is like a living being. It breathes in and out. As the blood goes round our body, so the waters of the earth go round, they rise as clouds, come down as rain and flow as streams back to the sea. The earth is like a living being, and living beings don't stand still, they move. Kepler's great quest was to find the answer to why God had made the cosmos as he did. Why had God placed the planets just at those distances from the sun, and not at other distances? And he wrote about this quest in his great book, *The Mysteries of the Cosmos.*

When war came to Austria, Kepler had to flee with his family, and came to Prague where he was appointed as Tycho's assistant. So these two great astronomers came together. But, as sometimes happens, the two did not get on. Tycho, the nobleman, was used to having many servants and assistants, who obeyed his instructions without question, and he treated Kepler like another of his assistants. Kepler who had been a university professor and was known for his writing throughout Europe felt he should be treated more as an equal, and not be left to eat at the servants' table. It also did not help that Kepler was a firm believer in Copernicus' system, while Tycho preferred his own.

However, the two had a grudging respect for each other.

For Tycho was the patient and accurate observer of the stars, while Kepler was a genius gifted with mathematical ability and insights.

When Kepler examined Tycho's accurate observations of the movement of the planets he found some irritating differences between the actual, observed position of the planet as Tycho had recorded, and the calculated position using the Copernican system. Kepler spent many years calculating. Only some years after Tycho Brahe had died, did Kepler at long last hit upon another idea. Perhaps things would work out more accurately if the planets and the earth were going round the sun not in a circle, but in an almost circular *ellipse.*

When Kepler worked out the paths of the planets and of earth round the sun as an ellipse, he found the planets really were in the places he had worked out. This is the world-picture which astronomers still have today: the moon moves round the earth in an ellipse, and the earth and the other planets move round the sun in ellipses.

The earth spins round on its axis, and we call this spin one day of 24 hours. And the earth moves in an ellipse round the sun, and one complete ellipse we call a year.

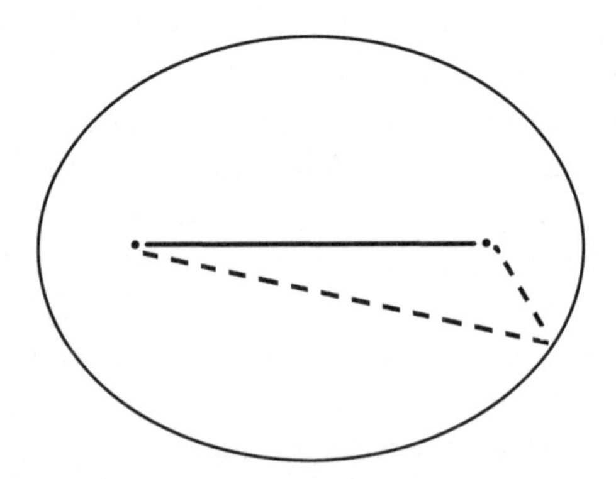

To draw an ellipse with a piece of string: knot a string in a loop, stick two pins into a paper, a small distance apart, and draw the ellipse by running a pencil around the string. The closer the pins are, the more the ellipse becomes like a circle; the further apart they are, the flatter the ellipse becomes.

But there is no real proof that the path of the earth is an ellipse: the calculations work out that way. And perhaps one day somebody will come with a quite new idea about the earth's movement or the sun's movement, and it will work out just as well. The only reason why the world-picture of Copernicus and Kepler is accepted today by all astronomers is that it is the easiest way for calculating where any planet will be, but that is not proof that the dance of the planets is as we imagine it.

38

Galileo and the Telescope

While Johannes Kepler was still working on his ellipses, a man in Holland had made a curious discovery. There is a story that it was not his own discovery, it was made by children. This man, called Hans Lippershey, was a master lens grinder who made spectacles. He saw two children playing with his lenses in his shop who said they could make the far away weather-vane appear upside-down but much closer.

A neighbouring spectacle-maker, Sacharias Jansen, seems to have come up with the same idea around the same time. No one has managed to find out which of the two spectacle-makers was really first or whether one stole the idea from another. But that was the beginning of the telescope and binoculars.

The Dutch spectacle makers were the first to discover the use of two lenses for telescopes, and it was also a Dutchman called Leeuwenhoek who made the first microscope. The Dutch wanted to keep their invention secret. At that time, the Dutch were at war against Spain and they thought it was a great advantage if they could observe with their telescopes what the enemies were doing far away, before the enemies could do so. They thought of the telescope only as something useful for their war. But, of course, such things cannot be kept secret, and rumours about the Dutch invention spread to other countries.

At that time in Italy there lived a man who is another of these heroes of thought, a hero of science. His name was Galileo Galilei, and he was deeply interested in astronomy. He had read the books of Copernicus, Tycho Brahe and Kepler, and he was quite certain that Copernicus and Kepler were right: that the earth moved round the sun. And when rumours of this Dutch invention of the telescope reached Galileo, he immediately realized that this new invention could be used for watching the stars. As he could not get any telescopes from the Dutch, who would not part with their secret, Galileo experimented and made his own telescope.

And so Galileo became the first man who looked at the stars, at the sun, moon and planets with a telescope. Imagine how he felt: he could see things no one had ever seen before; he not only saw more stars, but saw details of the moon and the planets no one had even thought of before. He made three great discoveries. First, he saw that certain dark bits on the moon were shadows cast by mountains — so the moon had mountains like the earth. Secondly, when looking at the planet Venus, he saw that it had phases like the moon. But his proudest discovery was that not only did the earth have a moon going round it, but that the planet Jupiter had moons. Galileo could see not just one, but four moons going round Jupiter. He was proud of his discoveries and he became famous as the first man who had seen the moons of Jupiter. But when Galileo wrote a book about his discoveries and defended the ideas of Copernicus and Kepler — that the earth travelled round the sun — it brought him into great trouble.

At that time monks and priests were the only people who had any learning. And if the Pope in Rome, the head of the Church, decided this or that was not good for people to know or to hear about, then that was the end of it. But now things were changing. Not only had Luther rebelled against the authority of the Catholic church in Rome, but people who were not monks or priests, like Tycho, Kepler, Galileo, came with all sorts of new ideas. The Pope in Rome and his Cardinals did not like it.

So Galileo was called to Rome, and the Pope was so powerful in those days that Galileo could not refuse. In Rome he was told that the Pope did not approve of what he had written. These ideas of Pythagoras, of Copernicus and Kepler — about the earth going round the sun — were not proven, and they also contradicted the word of the Holy Bible. (For instance, during a battle Joshua prayed to God to make the sun stand still, so that he could finish the battle before dark. How, asked the Cardinals, could God make the sun stand still if it was standing still already?)

Galileo could choose between publicly apologizing and admitting that this idea was a great mistake, or being imprisoned for the rest of his life. Galileo was by then an old man, and he had no wish to end his days locked up, and so, against his real belief, he apologized, called his book a great mistake, and was allowed to go.

From that time onwards he wrote and spoke no more about the ideas of Copernicus and Kepler, but, of course, he had not really changed his mind. So when science began it was not without danger to speak one's mind openly.

From the time of Galileo onwards more and more people looked at the stars through telescopes, and more and more discoveries have been made. Modern observatories have much bigger telescopes than Galileo had. Instead of lenses nowadays they use large concave mirrors. So you look down into a mirror at the sky, not up to the stars. At present (2011), the largest of these telescopes is in the Canary Islands, and it has a mirror more than 10 metres (33 ft) in diameter.

But there is something very curious about all the discoveries made with the help of these telescopes. All kinds of things are seen which no one had seen before, but the telescopes did not answer any questions, they only brought new riddles. Looking at the stars through a telescope has not proven or explained everything, it has only brought us new questions, new mysteries and new problems. It is quite possible that the priests of Babylonia who had no telescopes knew less but understood more about the cosmos than we do.

Of course astronomers do not look at the sun itself — this would ruin their eyesight. The telescope is directed at the sun and the picture of the sun caught on a screen. However, even with the largest telescope we do not see the sun itself. Just as the earth is surrounded by the atmosphere, so the sun, too, has a kind of atmosphere; but it is not air, it is shining light. As light in Greek is *photos,* this layer of light around the sun is called the *photosphere,* a sphere of light. This sphere of light around the sun is so bright that no human eye can see through it. So, no one has ever seen anything but the outer layer of light of the sun. What is inside this sphere of light, we do not know.

When Galileo first looked at the sun through a telescope (yes, he did ruin his eyesight, going almost blind in old age), he was puzzled. In the bright light he saw small dark spots. Galileo did not know what they were, and though we now know much more about these sunspots, we don't really know what they are. If we observe them for a few days, they appear to wander round the sun, indicating that the sun rotates about its axis, as earth does. But unlike the earth some parts of the sun rotate in about 21 days, other parts (away from the sun's equator) take longer, up to 26 days.

These mysterious sunspots come and go, none of them ever lasting more than a few months, and others appear. Sometimes there are more, sometimes there are less; but there seems to be a certain rhythm. Every eleven years there is a large number of sunspots, and then they become less. And — no one can tell you why — every eleven years here on earth is a specially good year for wine-growers.

Something somehow connected with sunspots is that when there are so many of them, there are days when radio reception is poorer than usual, the magnetic compasses of ships go awry, and the northern lights are more prominent.

Another thing which the telescopes show are enormous "flames" reaching out from the sun; they are called *solar flares.* Some of these mighty flares reach a height fifty times the size of our earth! They can only be seen during an eclipse or with a special telescope that blocks the light from the disc of the sun. At these times one can also see a wonderful halo of mother-of-pearl coloured light round the sun. It is called the *corona* of the sun (crown). But how the corona comes about, there are theories, but we don't really known.

The telescope has brought more and more questions about the sun.

39

Through the Telescope

The telescope brings new questions. Sunspots were discovered, but leave us with the question what sunspots really are. We don't know why their number increases every eleven years, or what causes the mighty solar flares or the beautiful corona. If we turn the telescope to our nearest neighbour in the cosmos, the moon, it is no different.

With the naked eye we see dark patches. We call it the man in the moon, in India and Africa they call it the hare in the moon. Looking through a telescope at the face of the moon we see mountains, and each mountain is a ring with a hole in it. These holes are called craters. Crater is the name given to the opening on top of a volcano, the opening where lava used to come out. Are the mountains on the moon volcanoes, are the round holes craters? We don't know. Most astronomers today say that these craters have been caused by meteorites falling on the moon. But it is just guesswork. All we do know is that these strange mountains have no snow-caps, no rivers, no clouds, and nothing grows on them. And we see these mountains so clearly in the telescope that there can be no air on the moon. But we don't really know what caused the craters or the mountains. The telescope has again brought us new questions but few answers.

Let us turn to Saturn. Saturn is not a very impressive sight for the naked eye, it is not very bright. But through the telescope, Saturn is the strangest sights of all the planets. There is a shiny globe (reflecting the sunlight, it has no light of its own), and around the globe there is a shining, broad ring. No other planet has such a ring. As a matter of fact, in a very strong telescope one can see that it is not really one ring but several concentric rings, one inside the other.

Again, these wonderful rings of Saturn pose all kinds of questions. For instance, why has only Saturn, of all the planets, such big rings? (Some of the other planets have very faint rings) What are the rings? One theory is that the rings are made up of thousands of tiny grains of dust which all go round Saturn, as our moon goes round the earth, but that is one theory. The strangest thing about these rings is that they form a very broad band around Saturn, with a diameter about 20 times that of the earth. But this enormous ring is as very, very thin, only about 20 metres. When Saturn is at a certain angle, we cannot see the rings at all. It is as if we were looking at the edge of a sheet of paper held level to our eyes. So, here again, the telescope shows us the rings of Saturn, but they are a mystery and a question.

Jupiter, the next planet, has several moons. Galileo saw four of them, another was discovered at the end of the nineteenth century. Four more were found in the first few years of the twentieth century, and another four were found before 1980. Now (2011) with space probes, astronomers reckon there are 63 moons around Jupiter. Looking at Jupiter through a telescope one sees darker and lighter bands going across the globe, called the belts of Jupiter. And these horizontal belts are forever shifting and changing. Astronomers think the shifting bands are different kinds of gas. Strangely, besides these shifting, changing belts, there is a large red spot, the Great Red Spot of Jupiter, but this red spot is always in the same place. We do not really know what this red spot is, or why it does not change or shift. Again, the telescope has brought us new questions.

The next planet is Mars, the "red" planet; when Mars is in the night-sky its red glow flashes like a ruby in the sky. Seen through a telescope Mars looks like a disc in a glowing orange colour. Astronomers say this orange colour is like some deserts here on earth, the yellow-orange desert sands on earth have this colour and so, perhaps, the whole of Mars is nothing but dead, dry desert. When, in the nineteenth century telescopes were powerful enough to see some detail of Mars, astronomers found a crisscross of lines which are of a grey-green colour, which were named canals. Later, with better telescopes, it was shown that these “canals” were optical illusions, like the dark spots we “see” where thick white lines intersect on a black background.

Another puzzle on Mars remains. On opposite sides of the orange disc of Mars there are white patches, reminding us, of course, of our Arctic and Antarctic. These white caps appear to be ice, like our poles. Mars has seasons, just as we have, Mars has summer and winter, only they take twice as long as our seasons. Here on earth when it is summer on the northern half it is winter in the southern half, in Australia, for instance. It is the same on Mars and we can see it in the telescope. When it is summer in the northern half of Mars, the white ice cap becomes smaller and smaller and at the same time the southern white cap grows bigger and bigger. After half a Mars-year things change round again, the small patch grows again and the big patch opposite shrinks. But the strange thing is that while here on earth our polar ice caps shrink or grow very little in the course of a year, on Mars they shrink and grow enormously. Again the telescope has presented us with more questions.

Now we come to the beautiful morning and evening star, Venus. The telescope can tell us nothing at all about Venus. Venus is surrounded by clouds, a complete, dense veil of clouds. The light of Venus is so dazzling white because the sun’s rays are reflected by this unbroken veil of clouds. But the telescope cannot let us see through clouds and so we know nothing about what is below the clouds on Venus. The goddess of beauty is veiled by clouds.

And Mercury is so small and so near to the sun, that the telescope cannot tell us much about Mercury either.

With the fixed stars we come again to mysteries. Most of the fixed stars are so far away that even with the strongest telescope they remain shining little dots of light. But the Milky Way which to our eyes looks like silvery dust, through the telescope is like a great river of countless stars. Some fixed stars through the telescopes are not points of light but are a shining cloud, called a *nebula.* A nebula is a shining cloud of some kind of gas. These nebulae have all kinds of shapes, but many have the beautiful shape of a spiral. One of the stars in Andromeda is such a shining, spiral cloud. These shining spiral nebulae are another mystery of the cosmos, presented to us through the telescope.

The telescope keeps giving us new questions and sets new challenges to science. The more we discover about the cosmos, the more answers we find, the more we become aware of new enigmas, of further questions.

40

Comets and Meteorites

There is another kind of strange wandering star, the comet. It is not often that a comet can be seen, but when it is visible in the sky it is such a strange sight that in the Middle Ages people used to be terrified when they saw them. When a comet first appears, it is simply noticed as a new star where none could be seen before. After a week or so, its position changes, and this new star not only grows brighter but also sprouts a kind of tail, a tail of light, which grows bigger and bigger. It is rather like a horse's tail, for it is finer and spread out towards the end. It is also curved, but it is a shining tail of light. After some months the tail fades, the head of the comet becomes fainter and then the whole comet disappears, and not even telescopes can find it.

Now astronomers were curious about two things: the tail of the comet, and its path. They noticed that the tail of every comet always pointed away from the sun. They also observed that at first no tail could be seen, and only when the comet came near to the sun did the tail appear, and then it disappeared as the comet went away from the sun. Astronomers think that the tail of the comet is a kind of air or gas, much finer and thinner than our air, so fine that the sun-rays can, like a wind, push it. The tail of the comet points away from the sun because it is pushed by the sun's rays. Astronomers think it is the sun's rays which make the tail shine, that make the gas glow.

Some comets come regularly, reappearing regularly in times between 20 years and 200 years. These comets move in a great ellipse round the sun, following the same laws as

Kepler discovered for the planets. But while the planets move in almost circular ellipses, comets move in very flat ellipses. Some comets move in such flat ellipses that astronomers can calculate it will take thousands of years or even much longer for them to return. But sometimes these comets appear a few times and then are never seen again. There are also comets which appeared suddenly and unexpectedly and have never been seen before. So we don't really know where they come from and what their path is once they go away.

There was a comet, Biela's Comet, which was observed for the first time in 1772. It came again a number of times in the nineteenth century and astronomers expected it to return in 1872, but they calculated that the comet would come right in the path of the earth and collide with the earth. Many people became terrified, believing the earth would be destroyed in the collision. But when 1872 came, there was no comet. All that happened was that there were more shooting stars, or meteorites, than usual. The comet was never seen again, and the meteorites were all that was left of this comet.

Meteorites are visitors from the cosmos. Perhaps all meteorites are bits of some ancient comets, but this is not certain. For bright comets are rather rare, but meteorites fall quite regularly at certain times of year. For instance, between August 11 and 13, there are a large number of meteorites, and again November 16–18 is a time of many meteorites falling. Astronomers think that there are swarms of meteorites in the path of the earth (like swarms of bees) and every time the earth moves into such a swarm we have a shower of shooting stars. But, sometimes — and no one knows why — there is a real downpour of meteorites. On the night of November 12–13, 1833, people in North America were treated to a strange show of firework; for nine hours the sky was ablaze with continuous flashes of falling meteorites.

The meteorite does not shine until it comes into the atmosphere surrounding the earth. Falling at a great speed, it heats up with the friction of the air, becoming white hot.

Most meteorites are completely burnt up high in the air but a number do reach the earth. The famous black stone, the Kaaba in Mecca which Muslims regard as holy, may well be a large meteorite. In 2007 a large meteorite fell in Peru near Lake Titicaca. It made a crater almost 5 metres deep, shattering the windows in a house 1 km away. And boiling water and foul-smelling gases started coming out of the crater. The meteorite had cracked the rocks allowing groundwater to rise together with trapped gases.

Although the most frequent meteorite falls are in August and November there are always meteorites falling. Some come to earth only as ash, as dust because they are burnt up, some come down in bigger lumps which are not completely burnt. Scientists have calculated that every year several hundred tons of meteorite come to earth. Over a thousand or ten thousand years, that is quite a lot, and one would expect the earth to grow bigger by thousands and thousands of tons over hundred thousands of years. But scientists have also worked out that this does not seem to have happened; the earth has not grown heavier or bigger. So somehow the earth gives back to the cosmos what it gets from the fall of meteorites.

The most interesting part of the meteorites is that some consist almost entirely of iron. Without iron we would have no machines, no cars, no railways or ships, no knives and no scissors. Not only that, but our blood contains iron. People who do not have enough iron in the blood are called anaemic, they are pale, feel weak and easily tire. It is the iron in the blood which gives us strength. And this iron which is in us, and in the earth, is also in the meteorites. These meteorites, these messengers from the cosmos, are nearly completely made of iron.

Where does the iron of the meteorites come from? Some scientists think that it comes from the sun, that the sunspots which have such an influence on the magnetism of the earth, are places where iron — so hot that it is a gas — is thrown out into the cosmos. And this iron-gas cools down and when it enters our atmosphere becomes the lump of iron which is either burnt in the air or is found here and there as meteorites. But that is not certain. Even iron from the cosmos, the meteorite, is a mystery. But when we see a shooting star, we can remember that it contains the same iron that gives strength to us through our blood.

41

The Atmosphere of the Earth

Compared with the great vastness of the cosmos, our own planet earth appears very small. There are millions of other suns in the cosmos: the stars. Every star is a sun, but so far away that the light of it is only a little sparkle in the sky. But astronomers who study the cosmos realize that our earth is a special place. Let us consider the planets of our solar system, of which earth is one. As a planet, the earth is much smaller than Jupiter, for instance. In fact Jupiter is bigger than all the other planets put together. But astronauts who have been in space and observed the moon, the stars and the sun from there have come to realize that the earth is a very special place.

James Irwin, an American who landed on the moon, said of the earth hanging in the blackness of space, "It diminished in size. Finally it shrank to the size of a marble, the most beautiful marble you can imagine. That beautiful, warm, living object looked so fragile, so delicate. Seeing this has to change a man, has to make a man appreciate the creation of God and the love of God."

The earth has something that no other planet has: an atmosphere with a clear blue sky, the beauty of sunrise or sunset, the ever changing weather, the winds which blow over the earth. There might be life somewhere else, there might be beings of quite another kind than man, but they will never see a rainbow, they will never see a blue sky or the glow of sunrise.

Whenever we look up to sun and stars, we see them through the atmosphere. Between us and the cosmos there is a great sphere surrounding the earth, the atmosphere. The atmosphere consists of air. Just as a fish lives at the bottom of the sea, so human beings, animals and plants live at the bottom of an ocean of air. People who climb high mountains or fly high in planes discover that the air gradually gets thinner the higher we go, and so they have to take air with them, they carry oxygen cylinders. Further down, we are always surrounded by this great ocean of air.

But this ocean is not only air, the atmosphere also contains water in the form of vapour. If we boil water it goes up as steam, it evaporates. But water also evaporates without boiling, as you can see when washing is hung out to dry, or when you spill some drops. The water goes away, it evaporates. This evaporation goes on all the time over land, but also over oceans, lakes, rivers, streams. Water is continually evaporating, and so the air of the atmosphere is filled with vapour. The air can hold a certain amount of water vapour, but if there is too much vapour, and then the vapour condenses back into water. If we have mist, fog, clouds, rain, then the air has more water than it can hold, and the air sends the water back to earth.

Warm air can hold more water vapour, and cold air holds less, just like a warm-hearted person receives visitors gladly, and a cold-hearted person will turn them away. There are always currents of warm air as well as currents of cold air in the atmosphere. Warm air rises and cold air falls, so even without the wind the air is continually moving. And where cold air and warm air come together, there is a battle, a clash, between the warm air and the cold air. What shows us where this battle is, are the clouds.

We sometimes see clouds that change their shape very quickly. This means that there is a very strong current of either warm or cold air. But sometimes a cloud formation does not change its shape much for hours, it does not even change its place, and we might think that nothing is happening. But, actually, a lot is going on in this unmoving cloud. From below warm vapour rises continuously, and when it meets cold air, the vapour turns into tiny, tiny droplets of water which hang in the air, they are too small to fall, they have hardly any weight at all. But as warm air reaches the droplets in the clouds and around the edges the droplets evaporate again and rise higher. So there is continually new vapour coming from below and becoming drops, and drops becoming vapour and floating away. The cloud shape remains the same, but all the time the drops which make it are changing.

To the west of Portugal there are islands called the Azores, and on one of these islands, Pico, there is a high mountain, and the top of this mountain is almost always surrounded by a cloud. This cloud has a quite regular shape, it does not change much, day and night, summer and winter. It keeps the same shape, stays in the same place, for years, even centuries. But the drops in the cloud change continually; new ones come from below, evaporate and disappear, floating away in warm air. The cloud is really like the course of a river, the riverbed remains the same, but the water is changing all the time.

At first glance it may seem that clouds can have all kinds of shapes, but there are scientists who study the shapes of clouds. (This science is very important for pilots of planes who must know from the look of a cloud what kind of air-currents they will encounter.) These scientists have found that there are four main kinds of clouds, All the clouds we see belong to one of the four groups, or are a mixture of two kinds.

Cumulus means "heap." Cumulus clouds look puffy like a sheep's fleece or like steam from a locomotive. Cumulus clouds have round forms, but the lower edge is always horizontal. The horizontal edge shows you exactly where the cold air begins. Here vapour becomes little drops.

Cirrus means "streak." They are wispy, feathery clouds, sometimes called mares' tails. Cirrus clouds are much higher than the cumulus clouds, so high that the droplets freeze, becoming tiny ice crystals.

Stratus means "layer." These are low layers of clouds. These low stratus clouds often veil mountain tops. Mountaineers don't like stratus clouds, for in such clouds they can't see where they are going. Mist and fog are really stratus clouds.

Nimbus just means "rain cloud." This type is always in a mixture. Nimbostratus is a shapeless grey cloud with steady rain or drizzle. Cumulonimbus is a very big, puffy cloud which can cause a thunderstorm.

Cirrus are the highest clouds, stratus are the lowest, cumulus are in-between. There are, of course, mixtures, for instance cumulostratus is the cloud we often see — large areas of lumpy clouds, often producing showers.

These clouds in the atmosphere — always changing — and the droplets in them continually evaporating and recondensing, are the carriers of life. For without clouds we would have no rain over the land, and without rain there would be no rivers, no lakes. The land would become desert, and all our precious life here on the earth would die out.

So having surveyed, the vast expanses of the cosmos, we come to see what a special place this earth of ours is.

Index

Also by Charles Kovacs

Class 4 (age 9–10)
Norse Mythology

Classes 4 and 5 (age 9–11)
The Human Being and the Animal World

Classes 5 and 6 (age 10–12)
Ancient Greece
Botany

Class 6 (age 11–12)
Ancient Rome

Classes 6 and 7 (age 11–13)
Geology and Astronomy

Class 7 (age 12–13)
The Age of Discovery

Classes 7 and 8 (age 12–14)
Muscles and Bones

Class 8 (age 13–14)
The Age of Revolution

Class 11 (age 16–17)
Parsifal and the Search for the Grail

The Apocalypse in Rudolf Steiner's Lecture Series
Christianity and the Ancient Mysteries
The Spiritual Background to Christian Festivals